Jason Grigsby

PROGRESSIVE WEB APPS

Publisher: Jeffrey Zeldman
Designer: Jason Santa Maria
Editor-in-Chief: Katel LeDû
Managing Editor: Lisa Maria Martin
Technical Editors: Aaron Gustafson and Alex Russell
Copyeditor: Katel LeDû
Proofreader: Mary van Ogtrop
Book Producer: Ron Bilodeau
Illustration Producer: Tyler Sticka

ISBN: 978-1-937557-72-0

A Book Apart
New York, New York
http://abookapart.com

TABLE OF CONTENTS

To my wife Dana,
who supported this book despite the fact that I swore I
would never write another book.
I'm sorry. I love you.

FOREWORD

THE IDEA OF "NATIVE APPS" always seemed like a regression. Walled gardens with terrible search, dubious security, and the endless tax of updates—it felt so 1990s. And yet, the web remained an objectively poor experience for mobile users. Responsive design helped, as did tools like Cordova and PhoneGap, but most sites weren't being built to address the real constraints of mobile.

We believed the web could be better, and set out to make it happen. It was a long shot, but by 2015, we succeeded in smuggling "installable web apps" into Chrome, with other browsers joining shortly thereafter. The technology had launched, but it didn't have a name. How could we succinctly communicate the range of complex technical properties that makes an online experience worth keeping? We found our answer in the moniker *progressive web apps*.

PWAs have always been motivated by user experience, not technology. Jason Grigsby was one of the first to grasp the implications of this UX shift and why browsers encourage it. Instead of focusing on the tedious aspects of the underlying technologies, Jason leads readers through the business and user consequences of building mobile-first—a journey on which PWAs are but one possible destination.

Knowing what content to make available when, how to give users control over their experience, and other oft-overlooked aspects are given the space they deserve, with thoughtful attention to detail. Each chapter crisply illustrates the business impact of delivering reliable, engaging, yet entirely "webby" experiences that earn user trust, as well as a coveted spot on homescreens and notification trays.

We could scarcely have hoped for a clearer distillation of the considerations, values, and benefits of progressive web apps. If the web wins, it will be the result of clear guides like this one lighting the way.

—Frances Berriman and Alex Russell

INTRODUCTION

YOU NEED A NATIVE APP. That's what we've been told repeatedly since Apple first announced the iPhone App Store.

And perhaps you do. Native apps can make sense depending on an organization's size and needs.

But what about potential customers who don't have your app? Or current customers on a desktop computer? What about people with limited space on their phones who delete apps to make room for other things? What is their experience like?

This is where *progressive web apps* (sometimes referred to as PWAs) shine. They combine the best features of the web with capabilities previously only available to native apps. Progressive web apps can be launched from an icon on the homescreen or in response to a push notification. They load nearly instantaneously and can be built to work offline (**FIG 0.1**).

Best of all, progressive web apps simply work. They are an enhancement to your website. No one needs to install anything to use a progressive web app. The first time someone visits your website, the features of a progressive web app are available immediately. No app stores (unless you want them). No gatekeepers. No barriers.

Early adopters of progressive web apps have seen significant returns on their investment. Housing.com increased conversions by 38 percent with a progressive web app (http://bkaprt.com/pwa/00-01/). Travel website Wego has seen 26 percent more visitors, a 95 percent increase in conversion, and a threefold increase in ad click-through rates (**FIG 0.2**) (http://bkaprt.com/pwa/00-02/). Lancôme's mobile sales increased 16 percent year over year after launching a progressive web app (http://bkaprt.com/pwa/00-03/).

Without knowing more about you and your organization, I can't say for certain whether you need a native app or not. But if you have a website—particularly one that is tied to revenue for your organization—you need a progressive web app.

The only questions that remain: What should your progressive web app do, and what should it look like? That's what this book is for.

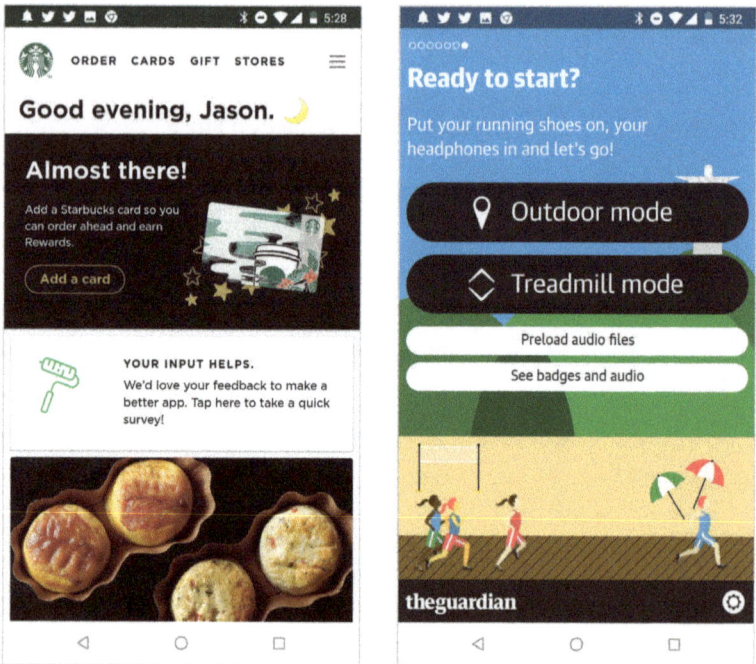

FIG 0.1: Starbucks' progressive web app (left) allows customers to manage their gift cards and pay for coffee even when they are offline. Rio Run, a running app by the *Guardian* (right), plays Brazilian music and provides information about historic venues along the Rio Olympics marathon route.

This book guides you through the decisions you'll need to make about your progressive web app, and how those decisions can impact the scope of your project. It will help you avoid common pitfalls and show how seemingly simple decisions can require far more functionality than their surface appearance implies.

While certain progressive web app features require JavaScript, there is very little code in this book. There are many resources for developers that describe how to build a progressive web app (many listed at the back of the book), but this isn't one of them.

This book was written for teams who are tasked with designing and building a progressive web app to understand what is

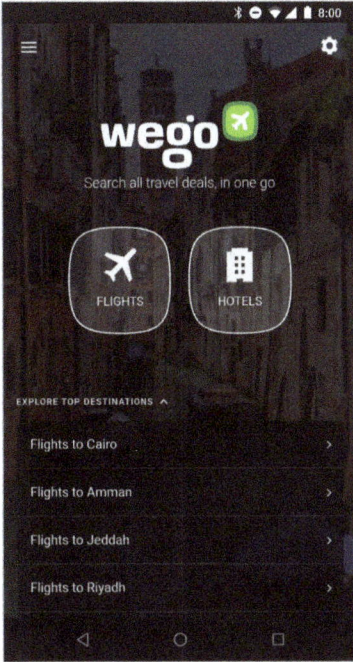

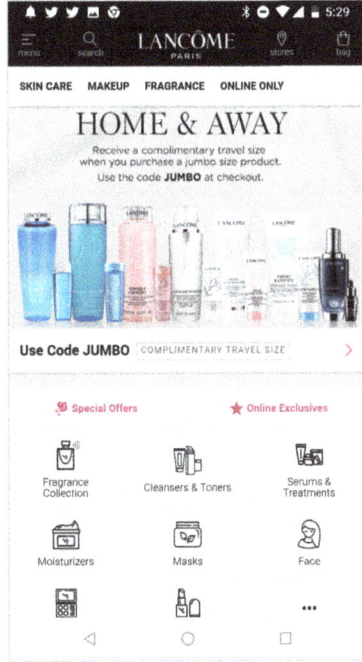

FIG 0.2: Wego (left) and Lancôme (right) are among several companies that report increased engagement, conversion, and revenue from progressive web apps.

being asked of them. It will help designers, product managers, and business unit owners gain a common understanding of what a progressive web app is and what features make sense for your organization. It will help your team define a roadmap, so your customers will benefit from progressive web app features as you complete them, instead of waiting for one big app release.

Progressive web apps represent much more than the discrete technologies that comprise them. They herald a new era where fast and immersive experiences are democratized. It's not surprising that some of the earliest adopters of progressive web apps have been in developing markets where storage is limited, connections are slow, and networks are metered.

We're on the verge of a web renaissance. I'm pleased you've decided to join us on the journey. Let's begin.

1

DEFINING PROGRESSIVE WEB APPS

LET'S GET THIS OUT OF THE WAY UP FRONT: the phrase *progressive web app* has become a bit of a buzzword. Its meaning often changes based on whom you are speaking with and what features they are most interested in.

And who can blame them? In 2015, Google's developer page on progressive web apps started with a list of ten characteristics, but was reduced to six just a year later (http://bkaprt.com/pwa/01-02/). Later the page added three characteristics—Reliable, Fast, and Engaging—none of which were in the original list of ten. Not content with that definition, at the 2017 Chrome Dev Summit, Google added "Integrated" to the list, to create the acronym "F.I.R.E." (http://bkaprt.com/pwa/01-03/).

When Google, one of the main proponents of progressive web apps, has changed its definition multiple times, it's no surprise that people are confused about what a progressive web app really is. And while F.I.R.E. is a great acronym, I'm not sure "Fast, Integrated, Reliable, and Engaging" offers much clarification.

Evolving definitions are fairly typical for new technologies. Back when HTML5 was considered buzzworthy, people would often talk enthusiastically about HTML5 features that were technically part of CSS3. Similarly, people often ascribe to progressive web apps features that are not technically part of the "official" definition.

Some of this fuzziness is intentional. Frances Berriman, who, along with Alex Russell, coined the phrase *progressive web apps*, said of the name, "It's marketing, just like HTML5 had very little to do with actual HTML. PWAs are just a bunch of technologies with a zingy-new brandname that keeps the open web going a bit longer" (http://bkaprt.com/pwa/01-04/).

So there's hype around progressive web apps—but you can use it to your advantage. When people get excited about progressive web apps, that's an opportunity to open minds about what is possible on the modern web. Organizations that pursue progressive web apps often end up reinvesting in the web, addressing longstanding issues like performance and user engagement. What Berriman and Russell did in naming a developing trend is similar to what Ethan Marcotte did for responsive web design: they gave everyone something to get excited about.

An amorphous definition can be helpful when drumming up interest in creating a progressive web app. But when your team sits down to start planning your PWA, you'll need a more concrete definition of what to build.

THE ORIGINAL DEFINITION

For many years, companies have tried to enable developers to create app-like experiences using web technology. Products ranging from Adobe AIR and PhoneGap to Windows Hosted Apps and Electron have sought to leverage the power of the web to create apps that exist outside the browser. Many apps have been created using these technologies, but each suffered the same problem: while they were built *using* the web, they weren't *of* the web.

In order to provide that app-like experience, these technologies were forced to give up two of the most powerful features of the web: the ability to link to anything, and the ability to run in any browser on any device. The tradeoffs necessary to make something feel like an app created something that was separated from the web.

In 2015, Berriman and Russell observed a new class of websites that were providing substantially better user experiences than traditional web applications. These new websites took advantage of a natural evolution in browser capabilities to create something that felt revolutionary: they escaped from the browser's tabs to live as their own apps, but retained the ubiquity and linkability that make the web what it is.

Berriman and Russell called these new websites *progressive web apps*, and documented nine defining characteristics (http://bkaprt.com/pwa/01-01/). These emergent sites were:

- **Responsive.** They leveraged responsive web design techniques to adapt the experience to whatever device or screen size a person might use.
- **Connectivity independent.** Portions of each website continued to function offline or on poor networks.
- **App-like.** They used app shell, animated transitions, and other touches to make it feel more "appy." (We'll talk about what that means more in Chapter 3.)
- **Fresh.** The app updated behind the scenes so it was always current.
- **Safe.** All traffic to the app was encrypted to prevent prying eyes.
- **Discoverable.** The sites declared that they were apps using a simple text file called a *manifest*, which search engines and browsers could look for to discover new apps.
- **Re-engageable.** They used push notifications to send alerts and bring users back into the app.
- **Installable.** Users could add an icon for the app to their homescreen.
- **Linkable.** The app could be shared easily, and accessed without having to download and install an app.

A final, key characteristic of progressive web apps is that they are built using *progressive enhancement*. Progressive enhancement is a web strategy that emphasizes creating a baseline experience that works everywhere, and then enhancing that experience on more capable devices.

What this means is that, while you will have a single progressive web app for all users, each individual's experience of that app may vary. A progressive web app can adapt based on the capabilities of the device and the permissions the user grants to the app, such as the permission to be on the homescreen.

Still, the definition of progressive web apps has been flexible from the beginning. Even Berriman and Russell deviated from their list of characteristics in that same post, when they highlighted Flipboard as an example, despite it not being responsive.

In fact, few progressive web apps meet all ten characteristics. Instead of thinking of this list as requirements, think of them as aspirations. Over time, our baseline expectations of progressive web apps will undoubtedly grow, but for now, you can have a successful progressive web app without checking every box on the list.

THE TECHNICAL DEFINITION

If Berriman and Russell's definition is aspirational, and Google's F.I.R.E. acronym is so broad as to be meaningless, how do we determine if something is a progressive web app or not?

Jeremy Keith proposed a more technical view in a 2017 blog post (http://bkaprt.com/pwa/01-05/). He defined progressive web apps by the inclusion of three technical features:

1. **HTTPS:** Progressive web apps must be served over a secure server using HTTPS. Most of the power in progressive web apps comes from service workers, which can only be used over HTTPS. If your current website is not using HTTPS, you need to move to HTTPS before you can do anything else.
2. **A service worker:** This powerful new technology allows web developers to intercept and control how a web browser handles its network requests and asset caching. With service workers, web developers can create reliably fast web pages and offline experiences.

3. **A web app manifest:** This short file ensures that progressive web apps are discoverable. It describes the name of the app, the start URL, icons, and all of the other details necessary to transform the website into an app.

A progressive web app may have more features than just these three, but it can't claim to be a progressive web app without them.

This nuts and bolts definition is useful for determining what qualifies as a progressive web app, but it's unlikely to inspire (or persuade your organization to open its purse strings). Instead, treat this definition as the minimum bar.

The vision of what a progressive web app can be is much broader. Think of progressive web apps as enabling experiences that previously required the development of a native application. Now, you can provide advanced features—like offline access and push notifications—to anyone who visits your website.

EMBRACE THE HYPE

I encourage you to embrace the ambiguous nature of progressive web apps. Ultimately, a strict definition matters much less than understanding what features make the most sense for your customers and your business. Your progressive web app will be unique to your website and your organization—and, as Jeremy Keith put it, there is value in tailoring how you talk about progressive web apps:

> If you're talking to the business people, tell them about the return on investment you get from Progressive Web Apps.
>
> If you're talking to the marketing people, tell them about the experiential benefits of Progressive Web Apps.
>
> But if you're talking to developers, tell them that a Progressive Web App is a website served over HTTPS with a service worker and manifest file. (http://bkaprt.com/pwa/01-05/)

When I was first introduced to the idea of progressive web apps, I admit I was unimpressed. From a technical perspective, there was nothing new here that I hadn't already been exposed to. I already believed every site should run HTTPS. I was familiar with service workers (and their troubled predecessor, app cache). Manifest files or packaging files had seemingly been around for as long as there had been browsers on mobile phones.

But what was routine to me represented exciting new possibilities for others. When I saw firsthand at a digital marketing conference just how many people were interested in the unique capabilities of PWAs, I changed my perspective. As Alex Russell wrote:

> It happens on the web from time to time that powerful technologies come to exist without the benefit of marketing departments or slick packaging. They linger and grow at the peripheries, becoming old-hat to a tiny group while remaining nearly invisible to everyone else. Until someone names them. (*http:// bkaprt.com/pwa/01-06/*)

In naming progressive web apps, Berriman and Russell opened the door for organizations to revisit what their website can and should do. And if you're wondering how to spark that conversation at your organization, look no further than the next chapter.

THE CASE FOR PWAS

NOW THAT YOU KNOW WHAT A progressive web app is, you're probably wondering if your organization would benefit from one. To determine if it makes sense for your organization, ask yourself two questions:

a) **Does your organization have a website?** If so, you would probably benefit from a progressive web app. This may sound flippant, but it's true: nearly every website should be a progressive web app, because they represent best practices for the web.

b) **Does your organization make money on your website via ecommerce, advertising, or some other method?** If so, you *definitely* need a progressive web app, because progressive web apps can have a significant impact on revenue.

This doesn't mean that your site needs to have every possible feature of progressive web apps. You may have no need to provide offline functionality, push notifications, or even the ability for people to install your website to their homescreen. You may only want the bare minimum: a secure site, a service

Comparing old mobile web to new mobile web

Time Spent > 5 minutes	User-generated Ad $	Ad Clickthroughs	Core Engagements
+40%	**+44%**	**+50%**	**+60%**

Comparing across web/native

Time Spent > 5 minutes	User-generated Ad $	Ad Clickthroughs	Core Engagements
+5%	**+2%**	**+0%**	**+2-3%**

FIG 2.1: Addy Osmani, an engineering manager for Google, wrote a case study about Pinterest's progressive web app, comparing it to both their old mobile website and their native app (http://bkaprt.com/pwa/02-03/).

worker to speed up the site, and a manifest file—things that benefit every website.

Of course, you may decide that your personal website or side project doesn't warrant the extra effort to make it into a progressive web app. That's understandable—and in the long run, even personal websites will gain progressive web app features when the underlying content management systems add support for them. For example, both Magento and WordPress have already announced their plans to bring progressive web apps to their respective platforms (http://bkaprt.com/pwa/02-01/, http://bkaprt.com/pwa/02-02/). Expect other platforms to follow suit.

But if you're running any kind of website that makes money for your organization, then it would behoove you to start planning for how to convert your website to a progressive web app. Companies that have deployed progressive web apps have seen increases in conversion, user engagement, sales, and advertising revenue. For example, Pinterest saw core engagement increase by 60 percent and user-generated ad revenue increase by 44 percent (**FIG 2.1**) (http://bkaprt.com/pwa/02-03/). West Elm saw a 15 percent increase in average time spent on their site and a 9 percent lift in revenue per visit (http://bkaprt.com/pwa/02-04/).

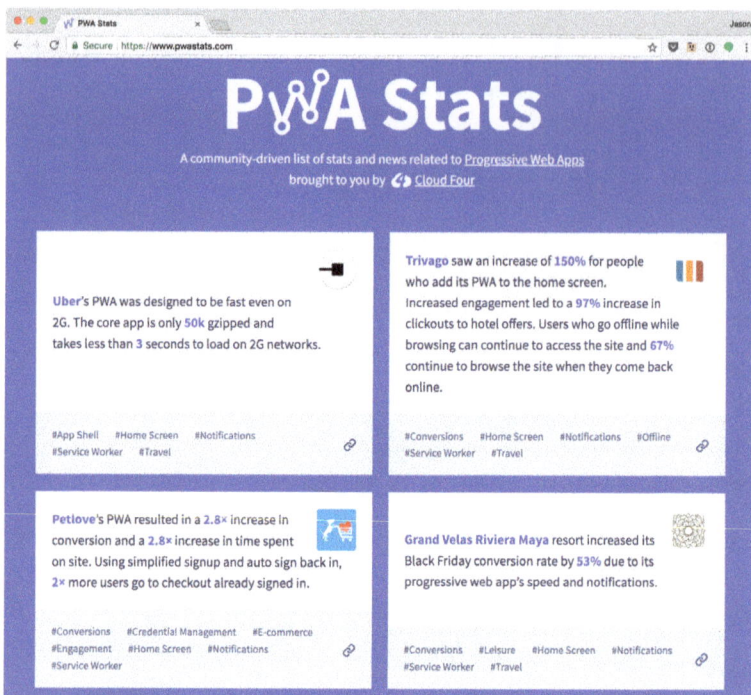

FIG 2.2: PWAstats.com collects statistics and stories documenting the impact of progressive web apps.

The success stories for progressive web apps are so abundant that my company, Cloud Four, started a website called PWA Stats (http://bkaprt.com/pwa/02-05/) to keep track of them (**FIG 2.2**). There's a good chance that we've collected a case study from an organization similar to yours that you can use to convince your coworkers that building a progressive web app makes sense.

And convincing them may be necessary. Despite the clear benefits of progressive web apps, many businesses still haven't converted—often because they simply don't know about PWAs yet. (So if you start building one now, you may get a jump on your competition!)

But there is also a lot of confusion about what progressive web apps are capable of, where they can be used, and how they relate to native apps. This confusion creates fear, uncertainty, and doubt (FUD) that slow the adoption of progressive web apps.

If you advocate for progressive web apps in your organization, you'll likely find some confusion and possibly even encounter some resistance. So let's equip you with arguments to cut through the FUD and convince your colleagues.

NATIVE APPS AND PWAS CAN COEXIST

If your organization already has a native app, stakeholders may balk at the idea of *also* having a progressive web app—especially since the main selling point of PWAs is to enable native app features and functionality.

It's tempting to view progressive web apps as competition to native apps—much of the press coverage has adopted this storyline. But the reality is that progressive web apps make sense irrespective of whether a company has a native app.

Set aside the "native versus web" debate, and focus on the experience you provide customers who interact with your organization via the web. Progressive web apps simply make sense on their own merits: they can help you reach more customers, secure your site, generate revenue, provide more reliable experiences, and notify users of updates—all as a complement to your native app.

Reach more customers

Not all of your current customers—and none of your potential customers—have your native app installed. Even your average customer is unlikely to have your app installed, and those customers who *do* have your app may still visit your site on a desktop computer.

Providing a better experience on the website itself will increase the chances that current and future customers will read your content or buy your products (or even download your native app!). A progressive web app can provide that better experience.

Despite what the tech press might have you believe, the mobile web is growing faster than native apps. comScore compared the top one thousand apps to the top one thousand mobile web properties and found that "mobile web audiences are almost 3x the size and growing 2x as fast as app audiences" (http://bkaprt.com/pwa/02-06/).

And while it's true that people spend more time in their favorite apps than they do on the web, you may have trouble convincing people to install your app in the first place. Over half of smartphone users in the United States don't download any apps in a typical month (http://bkaprt.com/pwa/02-07/).

Having a native app in an app store doesn't guarantee that people will install it. It costs a lot to advertise an app and convince people to try it. According to app marketing company Liftoff, the average cost to get someone to install an app is $4.12, and that shoots up to $8.21 per install if you want someone to create an account in your app (http://bkaprt.com/pwa/02-08/).

If you're lucky enough to get someone to install your app, the next hurdle is convincing them to continue to use it. When analyst Andrew Chen analyzed user retention data from 125 million mobile phones, he found that "the average app loses 77% of its DAUs [daily active users] within the first 3 days after the install. Within 30 days, it's lost 90% of DAUs. Within 90 days, it's over 95%" (**FIG 2.3**) (http://bkaprt.com/pwa/02-09/).

Progressive web apps don't have those same challenges. They're as easy for people to discover as your website is, because they *are* your website. And the features of a progressive web app are available immediately. There's no need to jump through the hoops of visiting an app store and downloading the app. Installation is fast: it happens in the background during the first site visit, and can literally be as simple as adding an icon to the homescreen.

As Alex Russell wrote in a 2017 Medium post:

Average Retention Curve for Android Apps

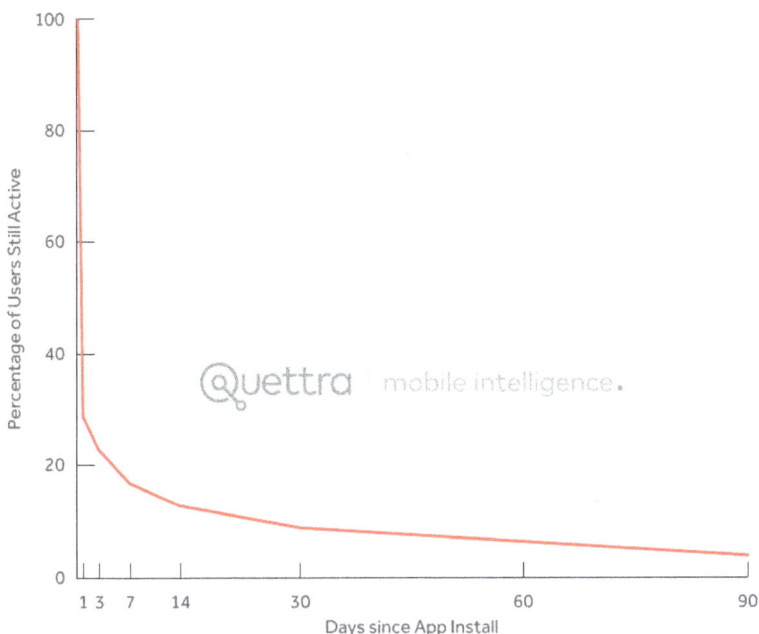

FIG 2.3: App loyalty remains a big issue for native apps. The average app loses over 95 percent of its daily active users within 90 days.

The friction of PWA installation is much lower. Our internal metrics at Google show that for similar volume of prompting for PWA banners and native app banners—the closest thing to an apples-to-apples comparison we can find—PWA banners convert 5-6x more often. More than half of users who chose to install a native app from these banners fail to complete installing the app whereas PWA installation is near-instant. (http://bkaprt.com/pwa/02-10/)

In short, a large and growing percentage of your customers interact with you on the web. Progressive web apps can lead to more revenue and engagement from more customers.

Secure your website

If you're collecting credit cards or private information, providing a secure website for your web visitors is a must. But even if your website doesn't handle sensitive data, it still makes sense to use HTTPS and provide a secure experience. Even seemingly innocuous web traffic can provide signals that can identify individuals and potentially compromise them. That's not to mention the concerns raised by revelations of government snooping.

It used to be that running a secure server was costly, confusing, and (seemingly) slower. Things have changed. SSL/TLS certificates used to cost hundreds of dollars, but now certificate provider Let's Encrypt gives them out for free (http://bkaprt. com/pwa/02-11/). Many hosting providers have integrated with certificate providers so you can set up HTTPS with a single click. And it turns out that HTTPS wasn't as slow as we thought it was (http://bkaprt.com/pwa/02-12/).

Websites on HTTPS can also move to a new version of HTTP called HTTP/2. The biggest benefit is that HTTP/2 is significantly faster that HTTP/1. For many hosting providers and content delivery networks (CDNs), the moment you move to HTTPS, you get HTTP/2 with no additional work.

If that wasn't enough incentive to move to HTTPS, browser makers are using a carrot-and-stick approach for pushing websites to make the change. For the stick, Chrome has started warning users when they enter data on a site that isn't running HTTPS. By the time you read this, Google plans to label all HTTP pages with a "Not secure" warning (**FIG 2.4**) (http:// bkaprt.com/pwa/02-13/). Other browsers will likely follow suit and start to flag sites that aren't encrypted to make sure users are aware that their data could be intercepted.

For the HTTPS carrot, browsers are starting to require HTTPS to use new features. If you want to utilize the latest and greatest web tech, you'll need to be running HTTPS (http:// bkaprt.com/pwa/02-15/). In fact, some features that used to work on nonsecure HTTP that are considered to contain sensitive data—for example, geolocation—are being restricted to HTTPS

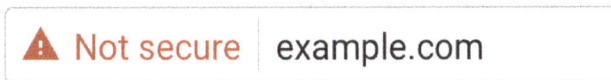

Eventual treatment of all
HTTP pages in Chrome:

⚠ Not secure | example.com

FIG 2.4: Google has announced its intention to label any website that isn't running HTTPS as not secure. Different warning styles will be rolled out over time, until the label reaches the final state shown here (http://bkaprt.com/pwa/02-14/).

now. On second thought, perhaps this is a bit of a stick as well. A carrot stick?

With all that in mind, it makes sense to set up a secure website for your visitors. You'll avoid scary nonsecure warnings. You'll get access to new browser features. You'll gain speed benefits from HTTP/2. And: you'll be setting yourself up for a progressive web app.

In order to use service workers, the core technology for progressive web apps, your website *must* be on HTTPS. So if you want to reap the rewards of all the PWA goodness, you need to do the work to make sure your foundation is secure.

Generate more revenue

There are numerous studies that show a connection between the speed of a website and the amount of time and money people are willing to spend on it. DoubleClick found that "53% of mobile site visits are abandoned if pages take longer than 3 seconds to load" (http://bkaprt.com/pwa/02-16/). Walmart found that for every 100 milliseconds of improvement to page load time, there was up to a one percent increase in incremental revenue (http://bkaprt.com/pwa/02-17/).

Providing a fast web experience makes a big difference to the bottom line. Unfortunately, the average load time for mobile websites is nineteen seconds on 3G connections (http://bkaprt.com/pwa/02-16/). That's where a progressive web app can help.

Progressive web apps use service workers to provide an exceptionally fast experience. Service workers allow developers to explicitly define what files the browser should store in its local cache and under what circumstances the browser should check for updates to the cached files. Files that are stored in the local cache can be accessed much more quickly than files that are retrieved from the network.

When someone requests a new page from a progressive web app, most of the files needed to render that page are already stored on the local device. This means that the page can load nearly instantaneously because all the browser needs to download is the incremental information needed for that page.

In many ways, this is the same thing that makes native apps so fast. When someone installs a native app, they download the files necessary to run the app ahead of time. After that occurs, the native app only has to retrieve any new data. Service workers allow the web to do something similar.

The impact of progressive web apps on performance can be astounding. For example, Tinder cut load times from 11.91 seconds to 4.69 seconds with their progressive web app—and it's 90 percent smaller than their native Android app. (http://bkaprt.com/pwa/02-18/). Hotel chain Treebo launched a progressive web app and saw a fourfold increase in conversion rates year-over-year; conversion rates for repeat users saw a threefold increase, and their median interactive time on mobile dropped to 1.5 seconds (http://bkaprt.com/pwa/02-19/).

Ensure network reliability

Mobile networks are flaky. One moment you're on a fast LTE connection, and the next you're slogging along at 2G speeds—or simply offline. We've all experienced situations like this. But our websites are still primarily built with an assumption that networks are reliable.

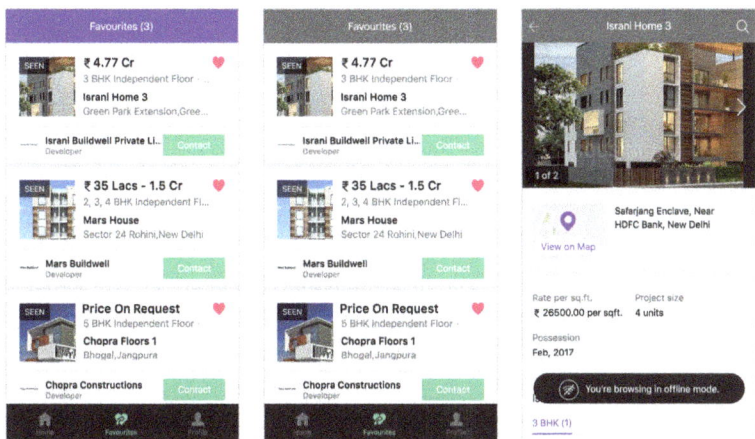

FIG 2.5: The header in housing.com's progressive web app changes from purple (left) to gray when offline (middle). Content the user has previously viewed or favorited is available offline (right), which is important for housing.com's home market in India, where network connectivity can be slow and unreliable.

With progressive web apps, you can create an app that continues to work when someone is offline. In fact, the technology used to create an offline experience is the same technology used to make web pages fast: service workers.

Remember, service workers allow us to explicitly tell the browser what to cache locally. We can expand what is stored locally—not only the assets needed to render the app, but also the content of pages—so that people can continue to view pages offline (**FIG 2.5**).

Using a service worker, we can even precache the shell of our application behind the scenes. This means that when someone visits a progressive web app for the first time, the whole application could be downloaded, stored in the cache, and ready for offline use without requiring the person to take any action to initiate it. For more on when precaching makes sense, see Chapter 5.

Keep users engaged

Push notifications are perhaps the best way to keep people engaged with an application. They prompt someone to return to an app with tantalizing nuggets of new information, from breaking news alerts to chat messages.

So why limit push notifications to those who install a native application? For instance, if you have a chat or social media application, wouldn't it be nice to notify people of new messages (FIG 2.6)?

Progressive web apps—specifically our friend the service worker—make push notifications possible for any website to use. Notifications aren't required for something to be a progressive web app, but they are often effective at increasing re-engagement and revenue:

- United eXtra Electronics saw a fourfold increase in re-engagement and a 100 percent increase in sales from users arriving via push notifications (http://bkaprt.com/pwa/02-20/).
- Notifications contributed to a 12 percent increase in recovered carts for Lancôme (http://bkaprt.com/pwa/02-21/).
- Classified ads company OLX saw a 250 percent increase in re-engagement using push notifications and a 146 percent higher click-through rate on ads (http://bkaprt.com/pwa/02-22/).
- Carnival Cruise Line garnered a 42 percent open rate from its push notifications. In addition to the 24 percent of mobile users who opted in to the notifications, 16 percent of desktop users opted in as well (http://bkaprt.com/pwa/02-23/).

We'll talk more about push notifications in Chapter 6. For now, it can be helpful to know that progressive web apps can send push notifications, just like a native app—which may help you make the case to your company.

Whether you have a native app or not, a progressive web app is probably right for you. Every step toward a progressive web app is a step toward a better website. Websites *should* be secure. They *should* be fast. They would be better if they were available offline and able to send notifications when necessary.

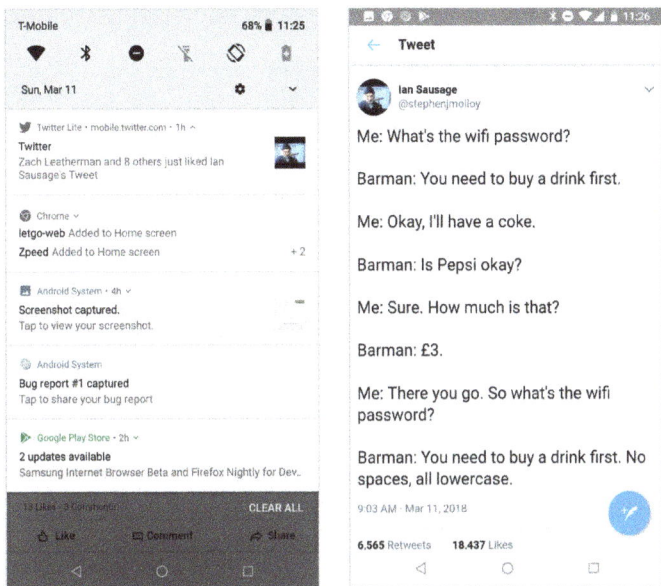

FIG 2.6: Twitter's progressive web app, Twitter Lite, sends the same notifications that its native app sends. They appear alongside other app notifications (left). Selecting one takes you directly to the referenced tweet in Twitter Lite (right).

For your customers who don't have or use your native app, providing them with a better website experience is an excellent move for your business. It's really that simple.

THE WEB CAN DO MORE

There are many, often misleading, articles debating the merits of progressive web apps versus native apps. Not only do these articles promote a false either-or dichotomy, but they also share an outdated understanding of what the web is capable of.

For example, a recent article in *Mobile Marketer* stated that progressive web apps "do not currently support all of the hardware components that traditional native apps support, includ-

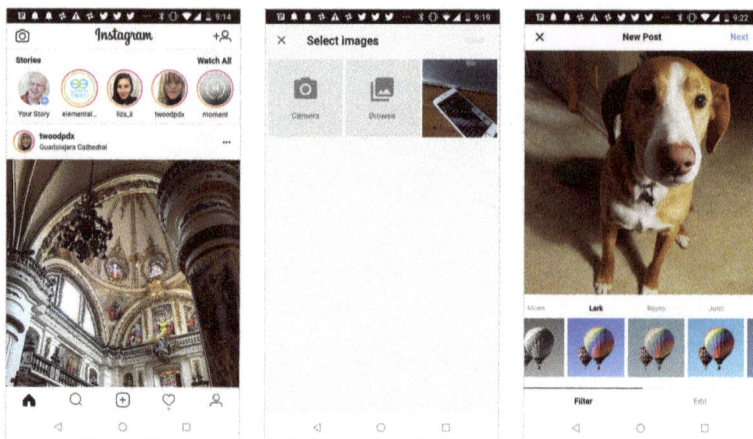

FIG 2.7: Instagram's progressive web app offers an experience that rivals its native app at a fraction of the size.

ing the camera, GPS, and fingerprint scanners on cell phones" (http://bkaprt.com/pwa/02-24/).

Yes, it's true that native apps generally gain access to device capabilities before the web does. But each one of those examples—camera, GPS, fingerprint scanners—can actually be accessed by web browsers in some fashion.

Access to GPS information via the Geolocation API has been available on the iPhone since 2009 (http://bkaprt.com/pwa/02-25/). You can hardly visit a website these days without being asked to share your location.

As for the camera, Instagram has built a progressive web app to help it expand its reach into developing countries. The progressive web app allows you to take photos and video and apply filters, and offers background sync via service workers so that if someone is offline, their photo will automatically upload when they are back online (FIG 2.7).

As for the fingerprint sensor, see Chapter 7 for information on the Payment Request API and how I accidentally bought a pair of socks with my fingerprint while trying to get a screenshot for this book. My credit card bill is proof that the web has access to the fingerprint sensor.

FIG 2.8: The phone's gyroscope and accelerometer let you send, receive, and track virtual paper planes across the world.

The point isn't to rebut this one article, but instead to illustrate how the web can often do more than we think it can. Web standards evolve, and gradually the web gains access to capabilities that used to be the sole domain of native apps. Most people don't have the time to keep track of what new standards have been released and adopted by browsers.

Another example of a progressive web app pushing the boundaries of what people think of as possible on the web is Paper Planes (http://bkaprt.com/pwa/02-26/). This app allows you to build a virtual paper airplane, stamp it with your location, and then send it across the globe with a flick of your wrist. You can also capture other users' planes in a virtual net and see where they have been. All of this is all possible with browser access to the mobile phone's gyroscope and accelerometer (**FIG 2.8**).

A lot has changed in the last few years. If it's been a while since the last time you researched whether or not the web supported some capability that would really make your progressive web app sing, you might want to check again. You may be pleasantly surprised at what the web can do these days.

BROWSER SUPPORT

One common early objection to progressive web apps was that Apple didn't support them. This is no longer the case—service workers shipped with iOS 11.3. This means that the most important features of progressive web apps—the performance and offline capabilities—are now available to Safari users.

With Safari's move to support service workers, the last major browser is now on board with progressive web app technology. But the concern about Safari's support for progressive web apps was always an overwrought consideration. There will always be people using older browsers that don't include progressive web app technology—but that shouldn't deter you from building a progressive web app.

Building a progressive web app can still benefit people even if their browser doesn't support all of the progressive web app features. Long before Apple committed to shipping service workers in Safari, multiple companies reported that their progressive web apps saw increases in conversion and engagement on iOS. AliExpress saw a 104 percent increase in conversion for new users across all browsers, and an 82 percent increase on iOS alone (http://bkaprt.com/pwa/02-27/). The *Washington Post*'s progressive web app saw a fivefold increase in user engagement, with minimal differences between Android and iOS (http://bkaprt.com/pwa/02-28/). And travel company Wego saw a 50 percent increase in conversion and a 35 percent increase in average session duration on iOS devices after building their progressive web app (http://bkaprt.com/pwa/00-02/).

If Safari didn't yet support progressive web apps, how can we explain how these companies see increases in conversion and engagement on iOS devices after building progressive web apps? While I can't say for certain, the most likely explanation is that progressive web apps are faster even on platforms that don't support service workers.

Organizations that want to take full advantage of progressive web apps will strive to build the fastest experience possible. Building that fast experience will mean tackling the things that slow down their current web experience—performance bugaboos like images that aren't optimized, large JavaScript frameworks, and excessive third-party scripts. Fixing these performance bottlenecks will benefit people on all browsers—even those browsers that don't "support" progressive web apps.

That's why I call progressive web apps a Trojan horse for website performance. For all of the showy features of progressive web apps—push notifications, offline usage, icons on the homescreen, app stores, etc.—the part of a progressive web app that will likely have the biggest impact for your users and your bottom line is performance. If the pursuit of those showy features causes your organization to build a faster experience, everyone wins.

A BETTER, FASTER WEB

Should your website be a progressive web app? The answer is almost certainly *yes*. Even if you don't think of your website as an "app," the core features of progressive web apps can benefit any website. Who wouldn't profit from a fast, secure, and reliable website?

From a business perspective, progressive web apps are winners. They have demonstrated the ability to increase conversion, sales, and ad revenue. They keep users engaged through push notifications. They allow organizations to reach a wider audience and maximize their website. Starting a progressive web app now can help you get ahead of your competition.

Once you cut through the FUD, it's clear that progressive web apps simply make sense. Progressive web apps unlock a world of possibilities—so many, in fact, that choosing what to do with your progressive web app may be your biggest challenge. The next few chapters will help you understand your options and their impact on your project.

3 MAKING IT FEEL LIKE AN APP

JEREMY KEITH ONCE OBSERVED, "Like obscenity and brunch, web apps can be described but not defined" (http://bkaprt.com/pwa/03-01/). But what does that mean for a team that begins work on a progressive web app?

Even if everyone in your organization believes that building a progressive web app is important, there is little chance that everyone has the same idea of what that app should look like, feel like, and do.

Creating a shared vision of what it means for your site to *feel* like an app is a crucial part of the process. How far you chase that dream will determine your scope, your features, your outcomes—everything.

So what does it mean to make something that is app-like? And—perhaps a better question—should that even be the goal?

"Being app-like" isn't a goal unto itself. Whether or not your progressive web app is more of an "app" or a "site" is not something your users are going to be thinking about as they interact with it.

It's better to focus on the characteristics of the experience that your users *will* notice. When we say "feel like an app," what we often mean is that the website should exhibit:

- a native look and feel,
- an immersive experience,
- speed and fluidity, and
- polish and personality.

The list of characteristics for you and your team may be different. But if you can translate your team's abstract sense of what makes something feel like an app into more concrete directions, then you can explore ways to achieve those characteristics.

Because I don't have your list handy, let's use my list for now and take a closer look at how we can make progressive web apps "feel" the way we want them to.

NATIVE LOOK AND FEEL

Owen Campbell-Moore, a former product manager for progressive web apps at Google, believes it is important to make progressive web apps fit in with their native app siblings, "since native apps have given users expectations around touch interactions and information hierarchy which are important to match to avoid creating a jarring experience" (http://bkaprt.com/pwa/03-02/).

There's something to be said for matching a user's expectations of their device. If someone launches your progressive web app from an icon on their Android phone, it would be nice if your app *feels* like an Android app. Or *feels* like an iOS app on an iPhone. Or a Windows app on Windows. You get the idea.

Fitting it visually with the device OS is certainly an option. For instance, you could use the system fonts provided by each operating system to make your app feel more like other apps on that platform. Geoff Graham maintains a reference on CSS-Tricks that describes how you can set the font in CSS to match the operating system font (http://bkaprt.com/pwa/03-03/):

```
body {
  font-family: -apple-system,BlinkMacS
  ystemFont,"Segoe UI",Roboto,Oxygen-
  Sans,Ubuntu,Cantarell,"Helvetica Neue",sans-serif
  }
```

The CSS Working Group has started to standardize on a new font-family value called system-ui that would always give you the operating system font (http://bkaprt.com/pwa/03-04/). While this new standard has been adopted by several browsers (http://bkaprt.com/pwa/03-05/), there have been issues with internationalization on Windows that have caused GitHub and Bootstrap to remove system-ui from their font list (http://bkaprt.com/pwa/03-06/). So be sure to test carefully if you use the new system-ui value.

If you want to go beyond fonts, you could adopt the platform's design language. Google's Material Design system has both official and unofficial ports to various CSS and JavaScript frameworks (http://bkaprt.com/pwa/03-07/). While there are no official web libraries from Apple to replicate the iOS design language on the web, there are plenty of third parties that try to mimic iOS look and feel using web technology. Microsoft's Fluent Design system is fairly new and, as of this writing, isn't licensed for web use.

While fitting in on the underlying platform makes some sense, it can be challenging to do well. Each platform has its own design language and guidelines that are extensively documented. How many platform design languages should you try to adapt to? What happens when the recommended design of one platform conflicts with the recommendations for another? What happens when the design language changes? Do you have to keep your progressive web app in sync with every new operating system release?

Don't forget, your progressive web app won't only be accessed via an icon on a mobile device. People will still be visiting your website in browsers and on other devices. Should we start making our widescreen responsive designs feel like Mac or Windows apps?

The uncanny valley

Even if you manage to keep on top of the evolving design language of multiple platforms, there's still a chance that your progressive web app will fall into the app equivalent of *the uncanny valley*—the idea that the closer robots, androids, and other humanoid objects get to appearing like real humans, the more real humans feel unsettled by them. We're much more likely to accept and feel warm towards a cartoony version of a human than we are one that looks nearly perfect, but feels off in subtle ways.

I suspect that something similar is true for web apps. The closer we get to trying to pass something off as a native app, the more glaring it will be when we don't match the design guidelines *exactly*. I believe we're much better off designing something that fits our own brand and provides an exceptional user experience across devices than we are trying to map the specific design aesthetics of any given platform.

We also don't know yet if people expect progressive web apps to behave similarly to native apps on their devices. Most people who add a progressive web app to their homescreen will do so after visiting the progressive web app in a browser. If they thought enough of the experience in their browser to add it to their homescreen, why would they expect that experience to morph into something different just because they are now launching the app from an icon on their device?

Unless there is a compelling business benefit, trying to make your progressive web app feel like a specific native platform is likely to be more pain than it is worth. Focus on making your app feel great no matter what operating system it's being used on.

IMMERSIVE EXPERIENCES

When someone visits your progressive web app in the friendly confines of their preferred browser, they'll see the site within the browser's *chrome*—that is, the window frame, toolbar, scroll bar, and other parts of the browser interface (FIG 3.1). In fact,

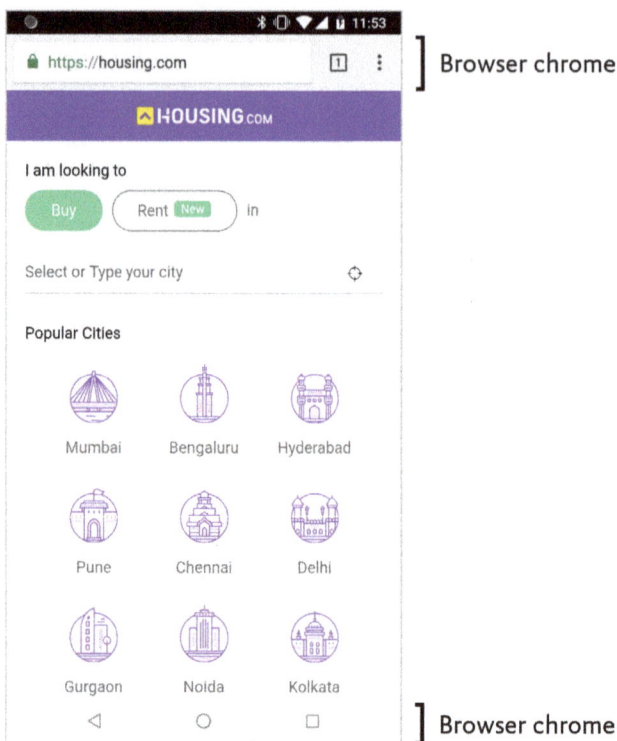

FIG 3.1: The chrome describes the parts of the user interface that surround web pages, including features like the address bar and the back button.

Google's browser was ironically named Chrome because a stated objective of the browser's design was to reduce the amount of browser chrome (http://bkaprt.com/pwa/03-08/).

As long as someone accesses your progressive web app inside a browser, you have no say over how much browser chrome they see. But the moment they add your progressive web app to their homescreen, you get to dictate how much of the browser user interface (UI) remains by declaring a display mode in the progressive web app's manifest file.

There are four display modes you can choose from (FIG 3.2) (http://bkaprt.com/pwa/03-09/):

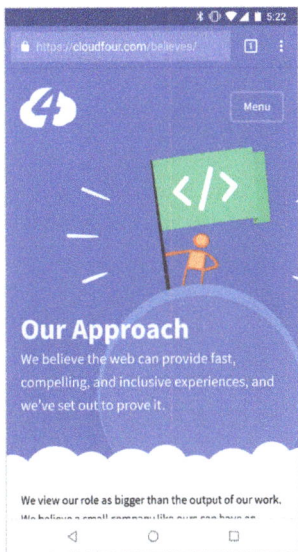

display-mode: browser

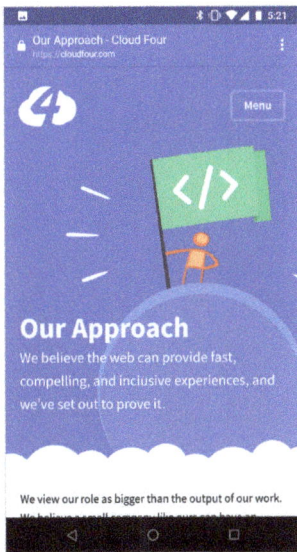

display-mode: minimal-ui

display-mode: standalone

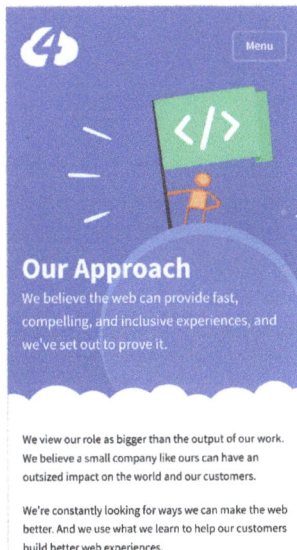

display-mode: fullscreen

FIG 3.2: Notice the differences in the Cloud Four website in Chrome with each of the four display modes.

- `browser`: Opens in a normal browser tab or new window.
- `minimal-ui`: Provides a minimal set of UI elements for controlling navigation and perhaps a way to view the URL of the page.
- `standalone`: The application will look like a standalone application. There may be system UI elements like the status bar and system back button.
- `fullscreen`: All available screen real estate belongs to your app.

Other than `fullscreen`, where nothing is visible except for the app itself, browsers have some discretion on how they implement each display mode and which display modes might satisfy the browser's installation criteria. Some browsers don't currently support all of the display modes. Because progressive web apps are fairly new, browser makers are still experimenting to see what features work best in each mode.

Removing browser chrome

Given the chance to finally break free of our browser shackles, it's hard to resist the `fullscreen` display mode. It's our chance to build the sort of rich, all-encompassing applications that used to require native code. Who wouldn't want to have the entire screen dedicated to showcasing your fabulous progressive web app?

But we should tread carefully when it comes to display modes. Those of us who build for the web have been spoiled by our browsers. It's easy to take for granted all of the things that browsers provide to us. As we move from the browser to `fullscreen` display mode, each step means the removal of functionality that our users rely on: the address bar, the back button, sharing options, printing, and much more.

Replicating these browser features inside a progressive web app is more difficult than it appears. Even the ubiquitous back button is complex. If you provide a back button, you will need to keep track of the browser's history and manage the application state to make sure that they go to the correct preceding

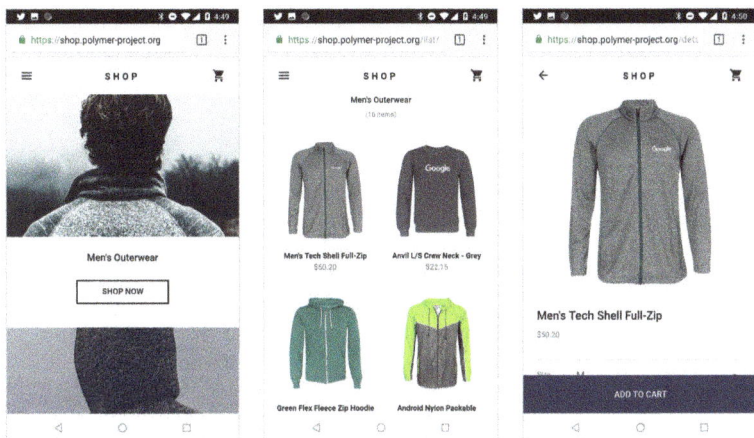

FIG 3.3: The Polymer demo site illustrates a typical ecommerce flow from home (left) to a category listing (middle) to a specific product view (right).

page. The History API can be used to manage and update the browser's history, but if you've never had to worry about tracking history because the browser took care of it for you, adding this functionality could be an unexpected hurdle.

If, as part of your progressive web app, you decide to switch to a single-page application structure (more on this and the app shell model soon), then you have an additional challenge of retaining scroll positions when implementing your back button. If you don't do this, then when someone hits your back button, they will be back at the top page instead of wherever they left off.

The back button in native mobile apps often supports a stricter navigation hierarchy than is typically present on the web. For example, a native ecommerce app might allow someone to drill down from the homescreen to a list of products, and then to a product detail page (**FIG 3.3**).

In this case, what the back button does is clear from the context: it takes you up the navigation hierarchy. And you know the hierarchy because you just successfully traversed it to reach the product detail page. People are accustomed to bouncing back

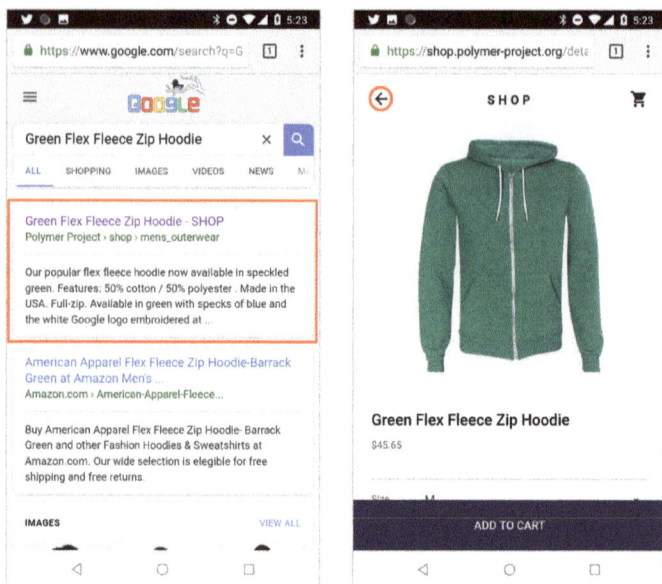

FIG 3.4: If someone follows a link in search results (left) to a product page in the Polymer Shop (right), where would users expect the back arrow in the body of the page to go?

and forth between the product detail and the product listing pages as they search for the right product.

Browsing the web is typically much less linear. It is just as likely that someone arrived on the product detail page from a search result or a tweeted link than it is that someone navigated through the hierarchy of categories and product listings. When that happens, where do they expect a back button provided by the progressive web app to go? Should it go up the navigation hierarchy to the product listing page like the native app does? Or should it go back to the search results (**FIG 3.4**)?

What happens if the progressive web app isn't being viewed on a mobile device? Where does the back button provided by the app go? Does it go to the same place that the back button provided by the browser goes? In the case of the Polymer demo ecommerce site, the back button provided by the progressive

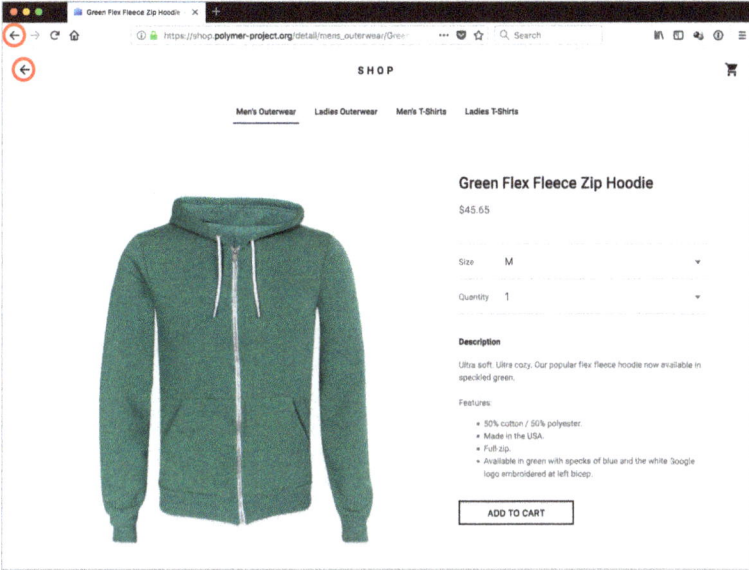

FIG 3.5: If someone has reached this page from search results, would the back arrow in the page go to the same location as the nearly identical back arrow provided by the browser?

web app goes to a different location—up the navigation hierarchy—than the back button provided by the browser, which goes back to the previous search result page (FIG 3.5).

Switching between the browser navigation and the in-app navigation in the Polymer demo creates funny results. For example, if you use the in-app back button to move up the navigation hierarchy to the product listing page and then use the browser's back button to get back to the product detail page, you'll now have arrows pointing in opposite directions—the in-app back button and a browser forward button—that both go to the product listing page. All roads lead to Rome indeed.

Detecting display modes

Each team that aspires to provide a fullscreen experience needs to ask itself what the experience should be when the progressive web app isn't viewed fullscreen. Does providing an in-app back button make sense if the browser back button is present?

Thankfully, there is a new media query that can be used to test for what display mode the progressive web app is currently being viewed in and then modify the layout accordingly. The following sample code could be used to display a back button only if the app were being viewed in fullscreen mode:

```
.backButton {
  display: none;
}
@media (display-mode: fullscreen) {
  .backButton {
    display: block;
  }
}
```

Note that in this example, the default experience assumes the visitor is viewing the page in a browser. More people are likely to view your progressive web app from inside a browser's chrome than a fullscreen experience.

Losing our URLs

When you declare your display mode to be either standalone or fullscreen, you lose the address bar and, with it, one of the cornerstones of the web: URLs.

That may not seem like much of a tradeoff to get access to more screen real estate. Who pays attention to ugly URLs anyways? But URLs are a key part of the web for good reason. They help users identify what site they are on. They help people know that the site is secure. They can be modified by users to get to other pages on a site.

Perhaps most importantly, URLs are what people share with others. If you find something you think someone else would like, you send them a URL. If you want to share an article on Facebook or Twitter, you share a URL. It becomes much more difficult for people to share the wonderful things they find on

FIG 3.6: In addition to two copy buttons, Bitly places the URL in an input field, which makes it easy to select and copy the URL.

your site if the address bar has been removed and there are no URLs available.

Just like the back button, if you go standalone or fullscreen, you will need to replicate the features that the browser's chrome provides inside your progressive web app.

One possible solution for sharing could be to add social network icons to your web page. Many websites feature a cacophony of social icons in an attempt to entice people to share their content—but there's little evidence that these icons are effective. One study found that only 0.2 percent of mobile users tap on social sharing buttons (http://bkaprt.com/pwa/03-10/). Additionally, if the JavaScript and images used to display the buttons come from a third-party source (like Facebook), then that third party knows what pages you have visited and can track your browsing experience (http://bkaprt.com/pwa/03-11/). In other words: these icons don't always work, and they sell out your visitors.

Another option is to make it easy for someone to access the URL and copy it. One UX pattern that works well is to place the URL in a form field that makes it easy to select and copy. Depending on the browser, you may also provide a copy button via the Clipboard API (**FIG 3.6**) (http://bkaprt.com/pwa/03-12/).

Chrome has added a notification on Android to help users copy and share URLs (**FIG 3.7**). However, because this is a *low-priority* notification, it doesn't draw much attention—it uses smaller text, doesn't buzz the device, and may get buried in the

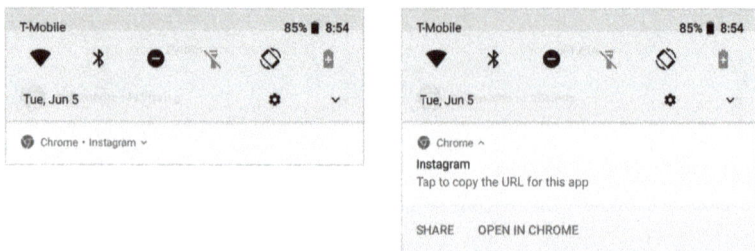

FIG 3.7: Chrome's new low-priority notification is designed to allow users access to copy and share the URL of a progressive web app in standalone or fullscreen mode—so long as they can find the notification. The notification (left) for Instagram's PWA is smaller than typical notifications, and doesn't explain what purpose it serves until the user opens it (right).

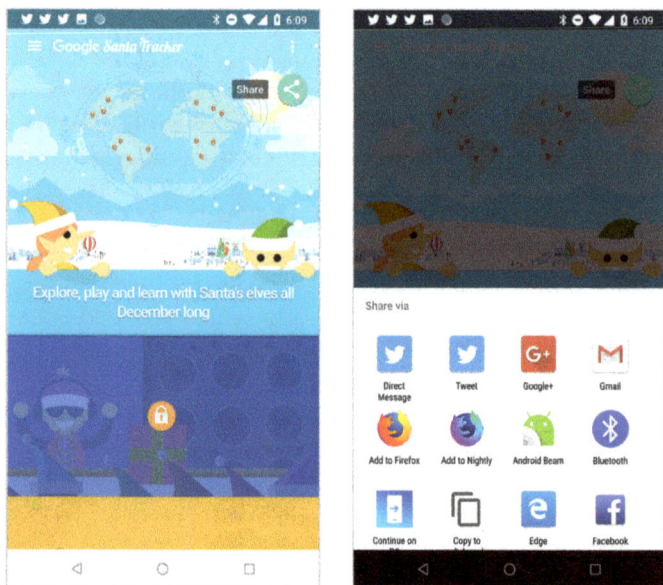

FIG 3.8: The 2017 version of Google's Santa Tracker (left) used the Web Share API. They saw a 20 percent increase in sharing (right) for users who had the API available to them (http://bkaprt.com/pwa/03-14/).

notification tray. Users will have to discover that this option exists, so you can't depend on them knowing about it.

In the future, it may be possible to use the Web Share API to connect to native sharing features on a given platform from within a web page. As of this writing, only Chrome and Opera currently support this standard (**FIG 3.8**) (http://bkaprt.com/pwa/03-13/).

Consider carefully whether the tradeoff of removing the sharing tools is worth it to gain the benefits of the additional screen real estate. You might find that the browser or minimal-ui display modes are better choices for your progressive web app. At a minimum, don't jump blindly into standalone or fullscreen modes without a plan for how users will navigate and share your content—or access any of the other features found in the browser's chrome.

It's remarkable how much impact one simple declaration—display: fullscreen or display: standalone—can have on the size of our endeavor and, ultimately, our user experiences.

Context continuum

It's easy to become fixated on what the experience is like when someone has installed our progressive web app. It's fun to imagine our users tapping on our icon on their homescreen and launching the immersive experience we've envisioned.

But the reality is that for most progressive web apps, the majority of people will visit them from within a browser. That's what makes progressive web apps powerful. People don't have to install anything before they can access and gain benefit from your progressive web app. You can build a single web experience that serves both highly engaged people who have your app installed as well as first-time visitors. This means everyone gets the best experience and you save on development and maintenance costs.

Therefore, it's important for us to keep in mind all of the varied ways in which someone will be interacting with our progressive web apps: in a mobile browser, on a desktop computer, in a web view inside a native app like Facebook, through an assistive device, and, hopefully, from an icon on their home-

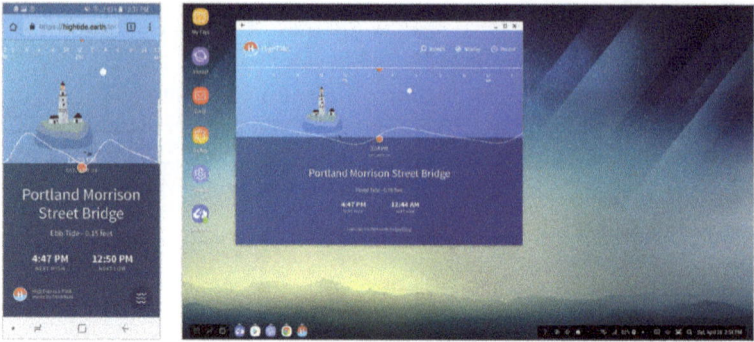

FIG 3.9: While most of the early attention on progressive web apps has focused on mobile devices, progressive web apps work (and in some cases can be installed) on desktop computers. The High Tide progressive web app can be displayed on a Samsung Galaxy S7 as a mobile view (left) or, when the S7 is connected via a dock, as a desktop view (right) (http://bkaprt.com/pwa/03-15/).

screen (FIG 3.9). There's a reason why the first characteristic of a progressive web app is that it's responsive.

Progressive web apps need to be considered from a continuum of browser contexts. A well-designed progressive web app will adapt to whatever the viewing context is.

SPEED AND FLUIDITY

One of the ways that Google suggests web developers make their website feel more "appy" is by using an app shell architecture. The shell refers to the common components necessary to build the user interface and power the app: navigation, header, footer, app logic—the things that should always be booted up when your app loads. Inside the shell is the content that changes from page to page (FIG 3.10).

Once the shell is separated from the content, it can be cached offline so that when someone opens the app, the shell loads instantly, no matter what the network conditions are like. All that the browser needs to retrieve to build that page, and any subsequent pages, is the new content. So not only does

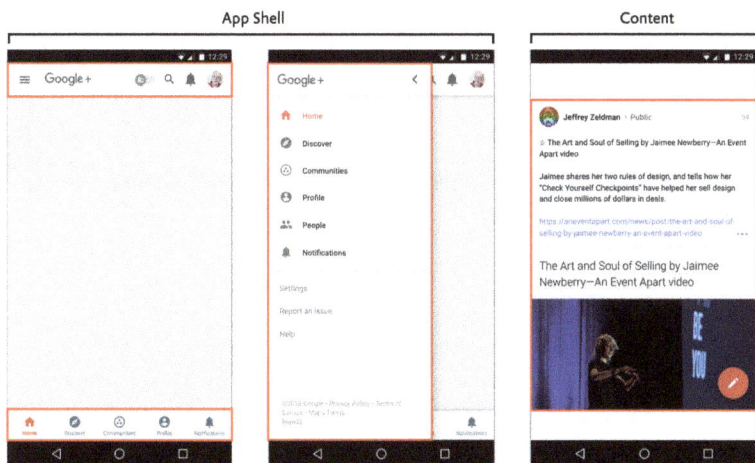

FIG 3.10: Google Plus uses an app shell for components that are repeated on every page and thus can be cached aggressively. The content of the page changes as someone interacts with the app.

the skeleton of the page load quickly, but the amount of data necessary for each page is reduced significantly, and you can animate transitions between pages.

The app shell model has benefits even for browsers that don't support service workers. The *perceived* speed of a web page matters more than the true speed. When app shell elements show up on the screen sooner, they create the perception of a faster browsing experience (**FIG 3.11**) (http://bkaprt.com/pwa/03-16/).

I believe this is one of the reasons why the *Washington Post* saw similar levels of engagement with its progressive web app test on iOS and Android. I conducted side-by-side tests of the *Washington Post*'s previous mobile web experience with its progressive web app. In one of my tests, I found that the progressive web app actually took more time to complete loading, but *felt* significantly faster than the old mobile website because the app shell model loaded the content that mattered sooner (http://bkaprt.com/pwa/03-17/).

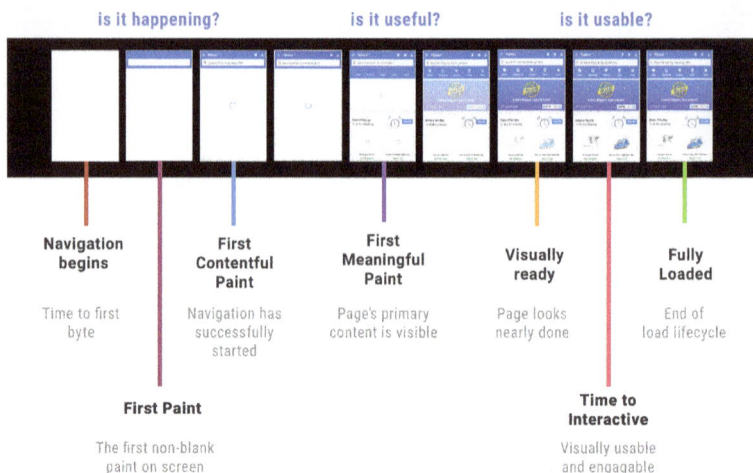

is it happening?	is it useful?	is it usable?

Navigation begins	First Contentful Paint	First Meaningful Paint	Visually ready	Fully Loaded
Time to first byte	Navigation has successfully started	Page's primary content is visible	Page looks nearly done	End of load lifecycle

First Paint

The first non-blank paint on screen

Time to Interactive

Visually usable and engagable

FIG 3.11: The perceived performance of a web page matters more than the total time it takes to download, as shown in this timeline created by Addy Osmani. Using the app shell model makes the page feel faster because the app shell can be displayed quickly while the content is being downloaded (http://bkaprt.com/pwa/03-16/).

Transitions between pages

In a traditional website architecture, each page loads as a unique item. You have no control over how the browser transitions between pages. But if you've built your progressive web app with an app shell, you gain the ability to control how your app transitions from page to page.

Controlling the way content comes into view can help users maintain a sense of the hierarchy in an application—particularly on a mobile device. When someone selects an item, you might slide the new content in from the right; when they select the back button, slide the content in from the left (**FIG 3.12**). This would reinforce the idea that moving to the right reveals more detailed information, while moving to the left backtracks up the app hierarchy (**FIG 3.13**).

Page transitions aren't limited to sliding from one direction or the other. Sarah Drasner recently described how to use web

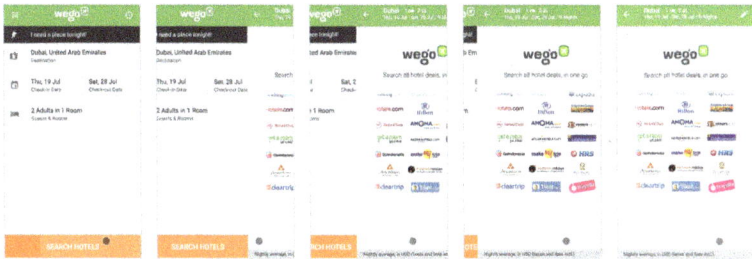

FIG 3.12: Wego slides content in from the right as you dig deeper into the app.

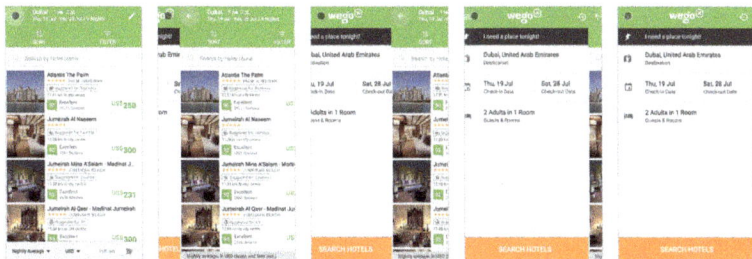

FIG 3.13: When you hit the back button in Wego's app, the screens slide in from the left, reinforcing the idea that you're backtracking your previous steps.

animation to create native-like transitions that keep one element of the page in focus as the page changes. On her demo site, the avatar remains visible and changes size and location as the user switches between pages in the application (**FIG 3.14**) (http://bkaprt.com/pwa/03-18/). Drasner explained:

> *Transitioning between two states can reduce cognitive load for your user, as when someone is scanning a page, they have to create a mental map of everything that's contained on it. ... Without these transitions, changes can be startling. They force the user to remap placement and even their understanding of the same element. It is for this reason that these effects become critical in an experience that helps the user feel at home and gather information quickly on the web. (http://bkaprt.com/pwa/03-19/)*

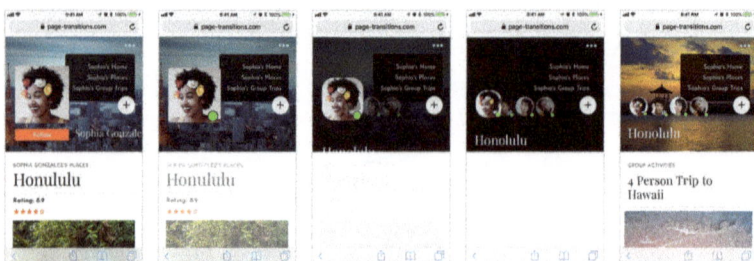

FIG 3.14: Sarah Drasner's demo site shows how focal-element animation can help a user stay oriented as they navigate. As the page transitions from Sophia's Places to Sophia's Group Trips, the photograph shrinks and becomes a circle, while the rest of the elements on the page fade out and are replaced with new content.

Smooth scrolling

As mentioned earlier, the perception of speed matters more than actual speed. If your progressive web app doesn't feel smooth, then it is likely to be perceived as slow, no matter how fast it truly is.

Pages that jump around while loading or sputter while scrolling frustrate users and diminish the polish of your progressive web app. Avoid late-loading images and scripts that modify the page layout. If items are going to load later, provide the browser with enough information to save space for the item in the layout.

Skeleton screens are one way of making pages feel fast and smooth while saving space in the layout for later loading items. Skeleton pages put placeholders on the page until the actual content loads (**FIG 3.15**). With service workers, these placeholders can be stored offline and loaded nearly instantaneously.

Pay close attention to how smooth and interactive your progressive web app feels. Be sure to test on mobile and lower-end devices to make sure that your app continues to perform smoothly on devices that have slower CPUs and less memory available. Too much JavaScript or too many layers will result in slow loads, sticky transitions, and poor scrolling behavior (http://bkaprt.com/pwa/03-20/).

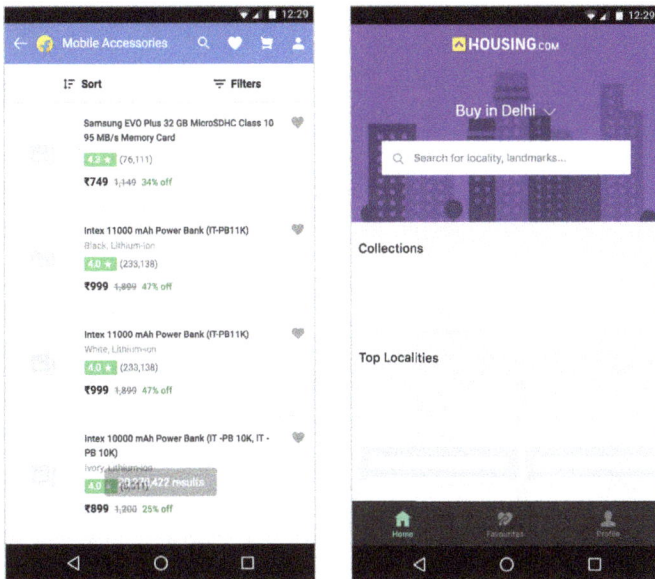

FIG 3.15: Flipkart (left) and housing.com (right) both utilize skeleton screens to increase perceived performance. The pages show progress by displaying placeholders where content and images will soon be displayed.

Over the years, there have been many attempts by browser makers and standards bodies to enable page transitions without adopting an app shell model. Unfortunately, none of these alternatives have gotten off the ground. If you want to control how your app transitions from page to page, you'll need to adopt an app shell model, or something similar like Barba.js (http://bkaprt.com/pwa/03-21/).

Single-page applications

The app shell model has obvious advantages, but it isn't a no-brainer decision to build a progressive web app using an app shell. Using app shell frequently implies that you're building a *single-page application* (sometimes referred to as a SPA). If your

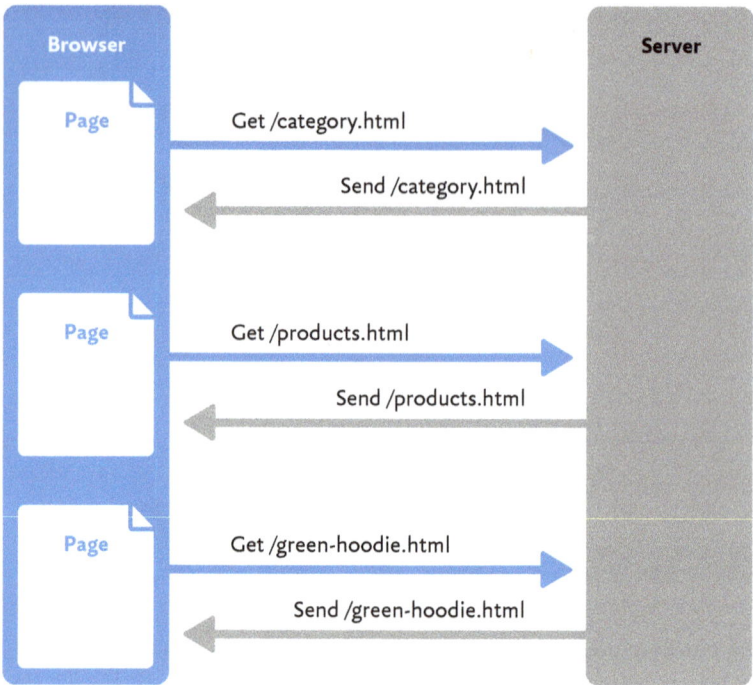

FIG 3.16: In a traditional web architecture, the browser requests a series of HTML documents. When the browser requests a new document, even if the document is on the same domain, the request starts fresh and the new page is rendered from scratch.

current website isn't using the single-page application model, moving to one can be a major undertaking.

Traditional websites were designed to support the idea that people download multiple HTML documents in sequence. A person might view the homepage, product listing, and product detail page in succession. For each page viewed, the browser would request and download a new HTML document and render an entirely new page. Even if the experience feels similar from page to page, the browser treats each document independently and replaces the previous document with a new one (**FIG 3.16**).

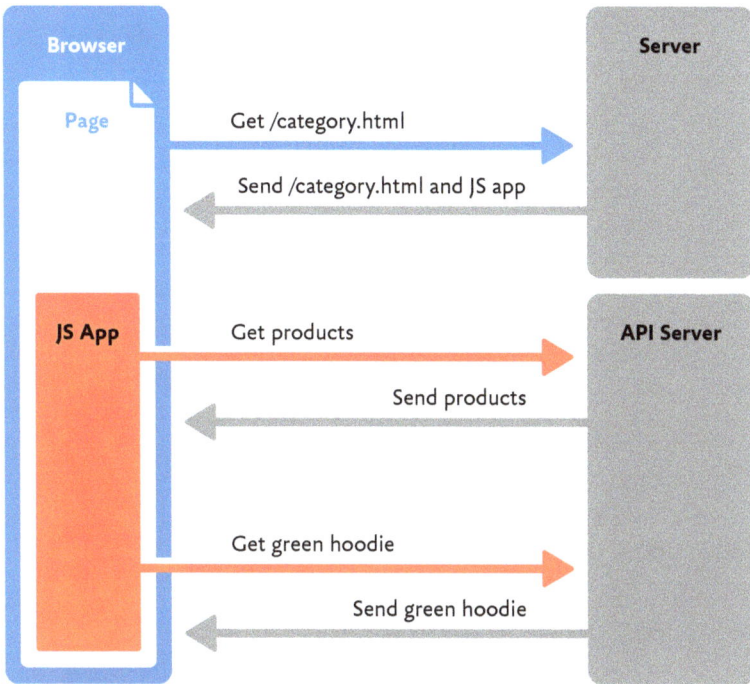

FIG 3.17: In a single-page application, the JavaScript application retrieves and displays new content within the same HTML document, rather than rendering separate pages with each new request.

With a single-page application, the first HTML document downloaded contains the information needed to power the application. Then, when any subsequent pages are requested, the application takes over from the browser and uses JavaScript to request the new content. The application then replaces the content of the current page with the new content (**FIG 3.17**).

This single-page application approach dovetails nicely with the app shell model because the app shell can stay in place after the JavaScript application boots up. But moving from traditional website architecture to a single-page application could require major changes in everything from the technology you use on your server to your team's skill sets and composition.

For example, to support a single-page application, the server must be able to respond to requests solely for content, rather than providing the full web page. This typically means that the server needs to provide application programming interfaces (APIs) that the application can consume. Ideally, these APIs will return content in a structured manner (such as a JSON file) that can be modified and manipulated by the JavaScript application. If your website has a traditional architecture, you may need to build those APIs before you can build a single-page application.

On the client-side, the complexity of the JavaScript required increases tremendously when you move to a single-page application. The JavaScript application must support history, routing, API calls, rendering content, and a host of other features that aren't necessary in a traditional architecture. Because of this complexity, this is a point at which developers often seek out JavaScript frameworks (like Vue.js, Svelte, Preact, or Polymer) that have already solved these problems.

Server-rendered JavaScript

If you *do* decide to build a single-page application using app shell, that doesn't mean you can ignore fundamental web principles such as progressive enhancement.

Single-page applications can speed up subsequent page loads after the application is up and running, but all of the application code still needs to be downloaded on that first page load. And if the application is complex, moving to a single-page application may slow down the site. You've likely encountered websites that seem like their pages *should* load quickly, but instead you're forced to watch a loading indicator (FIG 3.18). Often these sites have been built as a single-page application unnecessarily, forcing you to wait while the JavaScript application is started.

To avoid this slowdown, make sure that the server delivers a full HTML document containing all of the content on the first load, and then have the JavaScript application take over for any subsequent requests.

A popular way to do this is to use a JavaScript framework that supports server-side rendering. That way, the JavaScript code you write to render pages in the browser can also be

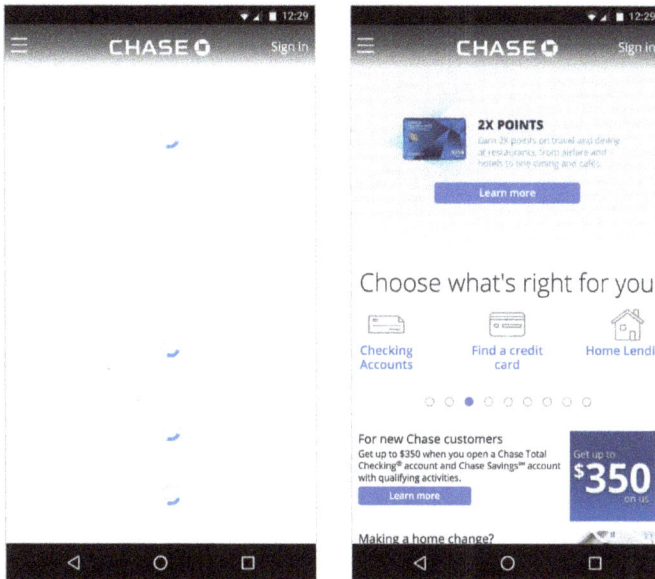

FIG 3.18: Chase Bank's website exhibits the telltale signs of a lack of server-side rendering. The app shell loads, and then multiple loading indicators appear (left) while the JavaScript application begins retrieving the content (right).

used to generate the initial HTML document sent by the server on a user's first visit (**FIG 3.19**). This idea of using the same JavaScript on the server and in the browser is referred to as *isomorphic JavaScript.*

Isomorphic JavaScript can provide significant time savings. For example, a product page template might need to connect to an API to retrieve information about a specific product. In a traditional web architecture, the code running on the server to connect to the API might be written in a language like Ruby or Python. But the same page might also need to connect to the same API from inside the browser, so all of the code would have to be rewritten in JavaScript. If, instead, you use a web server that understands JavaScript (such as Node.js), you can write code to connect to the API once and use it in both places.

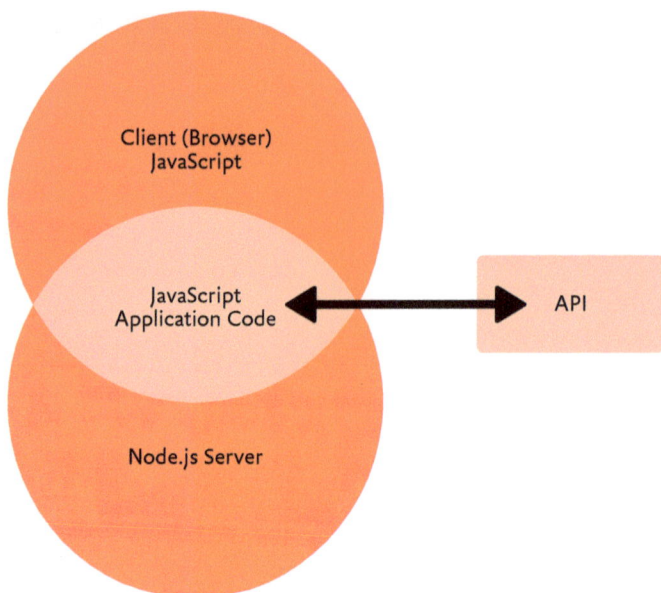

FIG 3.19: When you have a web server that understands JavaScript (like Node. js, one of the most popular for this purpose), you can use the same JavaScript application code on both the server and in the browser.

If you build a server-rendered, single-page application, take care to avoid silently breaking your user experience with too much JavaScript. Because JavaScript must be processed on the browser's main thread, the time it takes to download and process it delays interactivity for your users. As Alex Russell wrote, "JavaScript is the *single most* expensive part of any page in ways that are a function of both network capacity and device speed" (http://bkaprt.com/pwa/03-22/). Too frequently, single-page applications will download the content necessary to render the first paint of the page, but because JavaScript continues to process, the page itself is not interactive. Users can see the page, but they can't do anything with it— particularly on mobile devices where CPUs are slower.

A better approach for the performance of your progressive web app is what Google calls the *PRPL pattern* (http://bkaprt.com/pwa/03-23/):

- **Push** *critical resources for the initial URL route.*
- **Render** *initial route.*
- **Pre-cache** *remaining routes.*
- **Lazy-load** *and create remaining routes on demand.*

The pattern places an emphasis on making sure that the page is interactive as quickly as possible. Following the PRPL pattern allows you to defer loading JavaScript until much later when the service worker is installed. Doing so ensures that users will get always-interactive pixels on the first visit, and a better, more interactive experience on subsequent visits—all without suffering the lag of unknowably-large JavaScript download and evaluation.

In an ideal world, you have all of the APIs needed to support your progressive web app, your server supports isomorphic JavaScript, and your team is well versed on building performant single-page applications using the PRPL pattern.

But in my experience, few companies live in that ideal world. More likely, the move to an app shell model will require unforeseen, fundamental changes to your website's infrastructure. It isn't a change to be undertaken lightly, but there are substantial benefits to moving to an app shell model—particularly as part of a move towards more code reuse via isomorphic JavaScript and server-side rendering.

Shell-optional

Thankfully, app shell isn't required for a progressive web app. It's a common misconception that you need to use app shell if you want to build a progressive web app. Remember: all you need to convert a website into a progressive web app is to add HTTPS, a service worker, and a manifest file.

My company's site, cloudfour.com, was built on top of WordPress. We optimized it for performance before we made it into a progressive web app. After we added service workers, our site was so fast that we decided it didn't make sense to do the extra work to convert it to app shell.

If you feel strongly that you want to utilize app shell, but cannot do so immediately due to technical constraints, you may be able to use a library like Turbolinks, which attempts to provide a single-page application experience without requiring a rebuild of your entire website (http://bkaprt.com/pwa/03-24/). It works by fetching the full HTML page for subsequent requests and swapping out the body content (**FIG 3.20**).

Another way to support fast loading experiences without creating a single-page application is to take advantage of the new Streams API. This allows you to stream portions of your HTML and other files to the browser separately, which allows the browser to start building the page more quickly. For an example of how this works, check out Paul Kinlan's post on how he used the Streams API to create a feed reader application without an app shell (http://bkaprt.com/pwa/03-25/).

Bottom line: don't let the lack of an app shell prevent you from converting your current website into a progressive web app. Your visitors will benefit immediately from your efforts—app shell or no app shell.

POLISH AND PERSONALITY

When people talk about why they prefer apps, they often come back to intangible sentiments like "they just feel better." Apple talks about apps that "delight" users. In both cases, what this means is that the apps are providing a polished user interface imbued with personality. There's no reason why you can't provide the same with a progressive web app.

Native apps don't have a monopoly on UI polish. Yes, the tools that Apple, Google, and Microsoft provide can make it easier to create polished interfaces than starting from scratch in a blank HTML document, but there are plenty of slow and ugly native apps. What makes a great app is attention to detail in the design—particularly details like interactive feedback and animation.

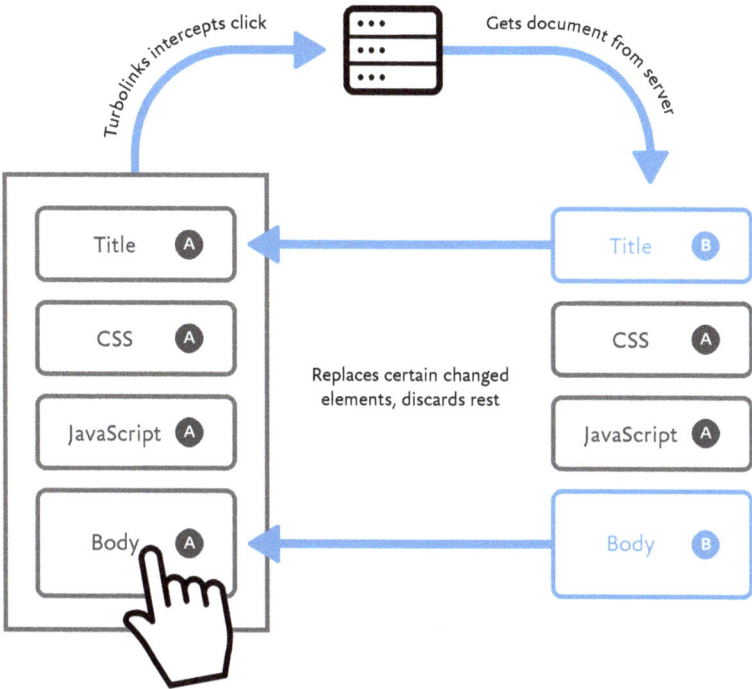

FIG 3.20: Turbolinks makes a traditional web page feel like a single-page application with app shell. It does this by intercepting requests for new pages, retrieving the new HTML document, and then selectively replacing parts of the old document with pieces of the new document.

Feedback

Apple's Human Interface Guidelines recommend providing "perceptible feedback in response to every user action" (http://bkaprt.com/pwa/03-26/). Immediate feedback—such as highlights, sounds, animations, and other responses to interaction—help users understand where they are, what they're doing, and how they can interact with the app.

Google's Material Design also places an emphasis on providing feedback with every interaction. For instance, selecting a button triggers a radial animation to let users know that their

FIG 3.21: Material Design uses a ripple effect when someone selects a button to provide instant feedback to the user that their action has been recognized (http://bkaprt.com/pwa/03-27/).

FIG 3.22: Stripe payment submission button provides feedback as it morphs from a regular button to a progress indicator and finally to a green checkmark.

selection has been registered (**FIG 3.21**). You can take inspiration from the feedback guidelines for native platforms, but don't limit yourself to them. Between CSS and SVG animations, nearly any animated feedback you might want to create is now possible.

Sound and vibration are commonly used in desktop sites and native apps. Facebook and LinkedIn use sound on the web to notify users when messages arrive. Twitter's native app also uses unobtrusive sound for notifications, and adds a vibration when users refresh their feeds—subtle touches that, unfortunately, aren't present in Twitter's progressive web app. Progressive web apps could easily include the same kinds of auditory and tactile feedback, but few do.

When designing feedback, try to convey not only that something has happened, but what, exactly, has happened. For example, Stripe's payment submission buttons immediately give visual feedback by morphing into a progress indicator, before becoming a checkbox to indicate that the payment has been submitted successfully (**FIG 3.22**).

Animation

One thing that commonly sets native apps apart from web apps is native's use of animation. For many years, browsers were incapable of displaying animations in a performant, stutter-free way. Now, however, web animations have reached a maturity level that make it possible to create the sort of polished interfaces we expect. Browser makers have moved as much animation as possible to the graphical processor unit (GPU) of the device so that the animation won't compete for resources with the rest of the page. These GPU-accelerated animations can be as smooth and rich as those available in native applications—assuming we take advantage of them.

Frankly, not enough websites *do*, which is a shame: animation is a way to make your progressive web app stand out. It's a powerful tool for providing feedback (as we've just seen), helping users understand how interfaces work, and enabling complex interactions on small screens. Best of all, it can play up your app's personality.

The opening and closing of menus, accordion interfaces, and carousels are places where some extra attention to the animation can make a difference. On a recent Cloud Four project, we redesigned the Wolfson Children's Hospital website, showcasing cheerful photographs of the children and their inspiring stories. To match that personality, my colleague Tyler Sticka created a playful animation for opening the menu (FIG 3.23). The navigation animation bounced in a childlike way, and incorporated the bright, primary colors of the Wolfson brand.

When working with the small canvas of mobile devices, animation becomes a necessary tool to help users understand the relationship between objects—particularly objects that are hidden or exist offscreen. On another Cloud Four project, we converted a Walmart Grocery interface that allowed customers to select a time slot for delivery. The widescreen version of the

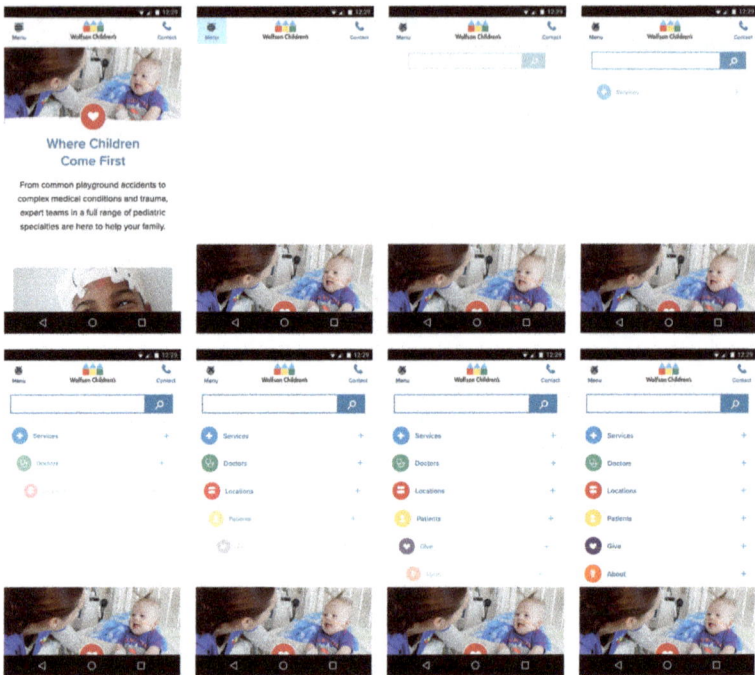

FIG 3.23: Wolfson Children's Hospital has a colorful and playful menu that animates quickly. The subtle touch reinforces the personality of the website.

chart showed several days of options, but when viewed on a mobile device, we only had room to show a single day (**FIG 3.24**).

Our initial solution was to let users move quickly to the next and previous days via buttons and a horizontally scrolling list. However, when we prototyped the interaction, we found it was a jarring experience: because the number of time slots varied each day, switching days caused the page to suddenly change in size and move sideways, leaving users confused as to what had just happened.

To help users connect the dots, we added animations (**FIG 3.25**):

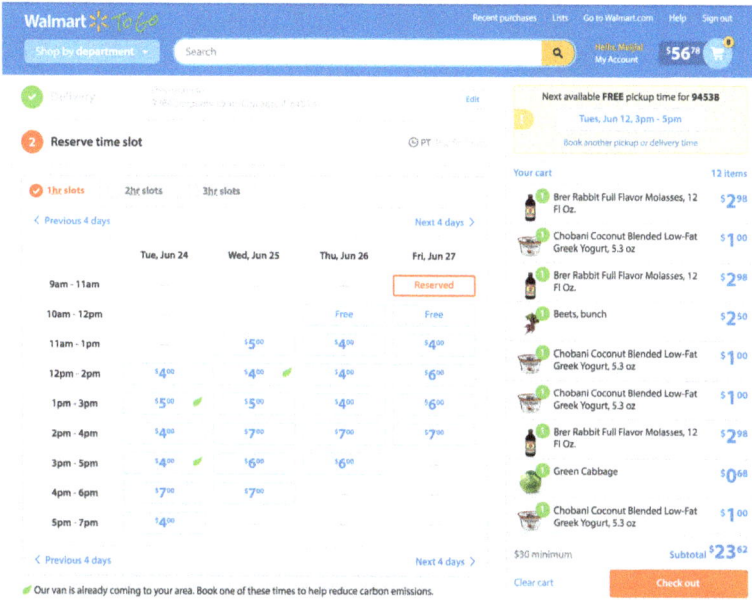

FIG 3.24: Walmart Grocery's checkout flow allowed users to select from multiple timeslots on multiple days on wide screens.

- After a user selects the button to go to another day, the page automatically scrolls up until the list of days is visible. By positioning the page at the top, the varying length of the content is less of an obstacle because the list of days is always in the same place. It provides a fixed object that gives users context for the animation.
- Then, the next day's time slots slide in from the right.
- As the next day slides into view, the list of days also slides from the right, and the green highlight moves from the one date to the next.
- If there are fewer available time slots, the page shrinks at a rate consistent with the time it takes to slide in that day's information.

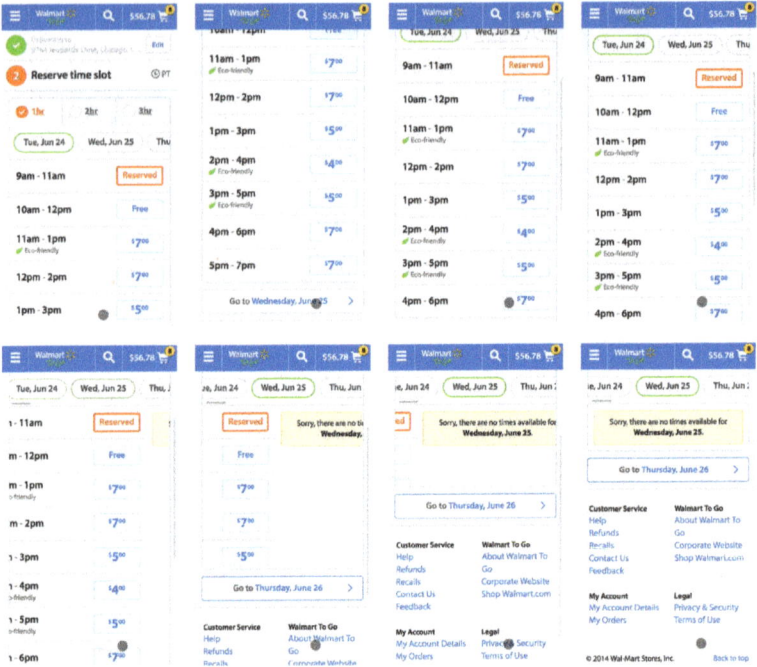

FIG 3.25: Walmart Grocery's time slot picker combines multiple animations to avoid a jarring experience.

FIG 3.26: Darin Senneff's animated login form demonstrates how animation done right can add personality and delight (http://bkaprt.com/pwa/03-28/).

All of these animations happen in a little over a second. People are unlikely to recognize the number of animations in use—nor the amount of testing and experimentation it took to find the right combination—but they would surely notice if the animation were missing.

Animation can also be fun and delight users, but when animation isn't tied to a specific UX goal, striking the right balance can be tough. Too many websites attempt to show off with animation, like the current fad of animating objects as they scroll into view. Often animation like this gets in the way and annoys users—undermining the design's objective.

However, it is possible to use animation that adds personality and doesn't detract from the user experience. For example, Darin Senneff created a demo login form featuring a friendly creature who reacts as users complete the form (**FIG 3.26**). As the user enters their email address, the creature follows along and smiles in a way that encourages the user. When they enter a password, the creature covers its eyes—but peeks if the Show checkbox is selected. This is the sort of touch that can make a progressive web app stand out from the competition.

As with page transitions and JavaScript enhancements, animations can generate lag if you're not careful. Browsers have become much better at supporting animations on a wide range of mobile devices, but be sure to test on underpowered devices to make sure your animations remain smooth. Stick to GPU-composited animation properties (http://bkaprt.com/pwa/03-29/), and reduce the number of layers that need to be composited (http://bkaprt.com/pwa/03-30/).

Animations should always be treated as enhancements, not requirements. Not all devices can support animation, including some types of assistive technology; many people, such as those with motion sensitivity or vestibular disorders, may prefer to turn animations off. Make sure that your designs can serve all users, even if the animation can't or won't be seen. For more information on the how and why behind responsible animation use, check out *Animation at Work* by Rachel Nabors.

BUILD AN EXCEPTIONAL EXPERIENCE

Remember that the more you push your progressive web app to "feel like an app," the more work it will be. That's not to say you shouldn't commit to that investment. Things like the app shell model, particularly when coupled with server-rendered JavaScript applications, can make a big difference in how a progressive web app feels, and can have additional benefits for your code reuse as well. Just recognize that there are surprises behind even simple things, like going fullscreen.

For your website, perhaps the "appy" parts of progressive web apps aren't necessary. You can still use progressive web app technology to speed up your site and provide a better user experience. There's nothing wrong with this approach.

If you want to build something on the other end of the site-to-app spectrum, be specific about what you want to accomplish. Being able to say what you want for your app—to feel faster, to be more immersive, to have more personality—will help you evaluate which solutions make the most sense.

In the end, whether or not something *feels* like an app or a website matters much less to our users than whether or not it is an exceptional experience. Build something that works for people, and they'll install it, use it, and won't care what you call it. But before they can install it, they need to be able to find it—so we'll look at making your app discoverable next.

4 INSTALLATION AND DISCOVERY

THE HOMESCREEN ICON IS BOTH the best and the worst feature of progressive web apps.

In the plus column, icons are attention-getters. Like the inclusion of *apps* in the name, if progressive web apps didn't offer the ability to be installed as an icon on the homescreen, I doubt people would be as excited about them.

But in the minus column, far too much emphasis is placed on the icon and the desire to have people "install" a progressive web app. Getting the icon on the homescreen ends up being the primary goal, instead of a byproduct of providing a great experience for your visitors.

What does it mean to "install" a progressive web app anyways? Our traditional notions of app installation don't map cleanly to progressive web apps. Progressive web apps don't rely on the user to download them and set them up—instead, the first time someone visits your site, the service worker is installed behind the scenes. This enables you do whatever you want with your progressive web app. Want to provide offline functionality? Do it. Want to send push notifications? Ask permission to do so.

You can have a fully functioning progressive web app that is only ever accessed from within the browser. The user may never put an icon on their homescreen, but they can still get all the benefits of your progressive web app. In fact, the majority of the people who use your progressive web app will likely never add the icon to their homescreen—they'll simply visit your website and utilize whatever features they find most useful.

And yet, getting the icon on the homescreen is treated like the holy grail, despite the fact that—as we saw in Chapter 2—the installation of a native app doesn't mean that people will return to or use that app.

Tunnel vision on icons causes people to miss seeing where their progressive web apps succeed. Even if no one ever added an icon to a homescreen, the progressive web app would still be worth the investment: everyone can benefit from a faster experience, offline behavior, push notifications, and other progressive web app features.

Of course, we want people to value our progressive web apps enough that they devote precious homescreen real estate to them. Therefore, we need to make sure that we've provided the browser with the information necessary to support the various ways that users might add an icon to their homescreen. The primary way we do this is through the manifest file.

WEB APP MANIFEST

When Apple released the iPhone, it added a handful of proprietary meta tags to describe what a web app should look like, what icon should be displayed, and how it would behave if someone added it to their homescreen. Because these meta tags were specific to Apple devices, Google and Microsoft eventually added their own meta tags. If a web developer supported all three platforms, the resulting HTML document would be littered with nearly two dozen meta tags (http://bkaprt.com/pwa/04-01/).

Manifest files solve this problem. The manifest file is a short, simple JSON document that describes the application, its icons, its background colors, and other details. When all browsers support manifest files, we'll be able to rip out all of the meta tags that we've added over the last decade.

Here is the manifest file from my company's website (http://bkaprt.com/pwa/04-02/):

```json
{
  "name": "Cloud Four",
  "short_name": "Cloud Four",
  "description": "We design and develop responsive
   websites and progressive web apps.",
  "dir": "ltr",
  "lang": "en",
  "icons": [
   {
     "src": "/android-chrome-192x192.png",
     "sizes": "192x192",
     "type": "image/png"
   },
   {
     "src": "/android-chrome-512x512.png",
     "sizes": "512x512",
     "type": "image/png"
   }
  ],
  "display": "standalone",
  "background_color": "#456BD9",
  "theme_color": "#456BD9",
  "orientation": "natural",
  "start_url": "/",
  "categories": ["business", "technology", "web"]
}
```

If you read the Web App Manifest specification, you'll see that the attributes in a manifest file are called *members*. There are many members listed in the spec, but let's look at the ones that are essential for most progressive web apps (http://bkaprt.com/pwa/04-03/).

Our manifest file starts by providing the basic details of our progressive web app:

```
"name": "Cloud Four",
"short_name": "Cloud Four",
"description": "We design and develop responsive
websites and progressive web apps."
```

The name can be used by app stores or browsers on startup screens and prompts. The short_name is typically used on the homescreen in conjunction with the icon. The description provides additional context for what the app is about.

Next, we can describe the language and text direction of the items in the manifest:

```
"dir": "ltr",
"lang": "en",
```

In this case, the text direction is set to left-to-right by the ltr value of the dir member. The language is English as represented by "lang": "en".

The icons member provides an array of possible icons that the browser can choose from, along with their sizes and type of image file:

```
"icons": [
  {
    "src": "/android-chrome-192x192.png",
    "sizes": "192x192",
    "type": "image/png"
  },
  {
    "src": "/android-chrome-512x512.png",
    "sizes": "512x512",
    "type": "image/png"
  }
],
```

As progressive web apps gain popularity and are used in different ways, the need for multiple icon sizes will increase. Be sure to check the latest manifest information to see what icon sizes are currently recommended.

Now we get to describe how we want the progressive web app to appear by setting display mode, colors, and orientation.

```
"display": "standalone",
"background_color": "#456BD9",
"theme_color": "#456BD9",
"orientation": "natural",
```

The display member declares the display mode that we talked about in Chapter 3. You can also declare an orientation preference, which can be useful for games or other apps that require a particular orientation. However, most progressive web apps should be responsive, which means the orientation can be omitted, or declared natural, as we've done here.

The background_color tells the browser what color to use on the startup splash screen that users will see when they launch your app. Each browser handles it differently, but typically the splash screen contains some combination of an icon and application name (or short name) set on the specified background color (FIG 4.1).

The theme_color is used by some browsers to tint the status bar, address bar, and other parts of the browser user interface (FIG 4.2).

Because browsers handle these color values and splash screens differently, it is important to test your app and make sure you're happy with how these settings look on different devices.

Next in our manifest file, we see:

```
"start_url": "/",
```

The start_url indicates what page should launch when someone opens your progressive web app. In this case, the start page is the homepage, which exists at the root or / path of our website. On other sites, the progressive web app might launch from a subdirectory.

You could also add URL parameters to the start URL:

```
"start_url": "/?utm_source=homescreen",
```

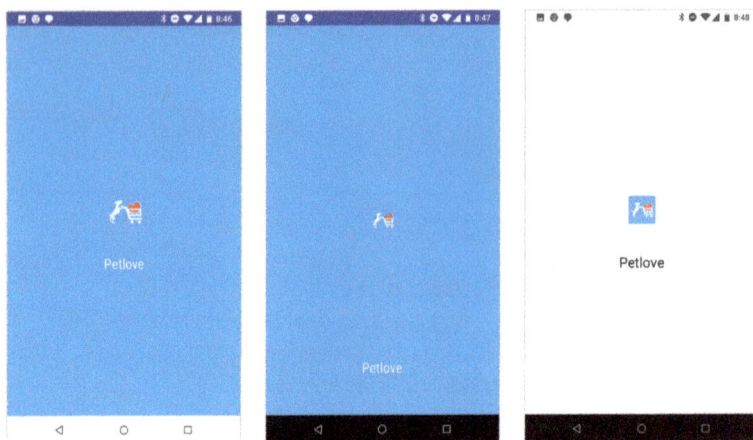

FIG 4.1: Each browser handles the splash screen differently. For instance, Chrome (left) has a fairly large icon, whereas Opera's icon (middle) is tiny. Samsung Internet (right) ignores the background_color declaration altogether—and Firefox doesn't display a splash screen at all. The lesson here is to test the startup screen in different browsers to make sure it is acceptable.

FIG 4.2: Lyft uses theme_color to tie the browser's user interface (left) more closely to its brand (right).

Using a start_url like that one would let you track in Google Analytics how many visits to your site launched from a homescreen icon.

The latest addition to the Web App Manifest specification is the categories member:

```
"categories": ["business", "technology", "web"]
```

This is intended to be used by app stores to categorize your app. As the addition of categories shows, the Web App Manifest is an evolving specification. It's worth referencing the

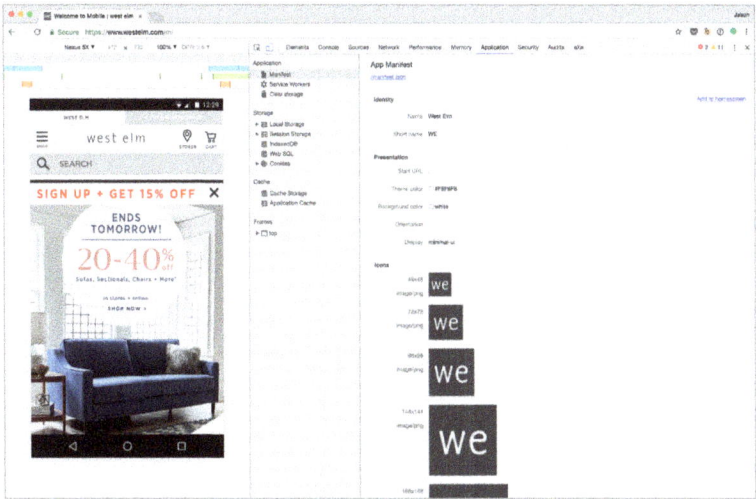

FIG 4.3: Chrome DevTools displays the details of a site's manifest file on the Application tab.

specification itself to see if there are any additional properties that your progressive web app would benefit from (http://bkaprt.com/pwa/04-03/).

The final step before our manifest is complete is to point to it from our HTML files. We do this by adding a link element inside the head. The manifest file link should be an absolute path, or you risk the manifest file breaking depending on the web page's location.

```
<link rel="manifest" href="/manifest.json">
```

This code will tell browsers, search engines, and anyone else who is interested where they can find the manifest that describes our progressive web app.

You can test your manifest file in the Application tab of Chrome DevTools (FIG 4.3). You can also use the Web Manifest Validator (http://bkaprt.com/pwa/04-04/), or Lighthouse, an open-source tool which is integrated into the Audits tab in Chrome

DevTools (http://bkaprt.com/pwa/04-05/). Lighthouse will also tell you if your app passes Chrome's installation criteria.

BANNERS AND BADGES

What browsers do with the information in the manifest file is up to them. One of the primary uses has been to help users discover that they've visited a progressive web app and notify them that they can add the app to their homescreen. There are two fundamental ways that browsers have been notifying users: banners and badges.

Getting prompted

Chrome and Opera have both implemented banner prompts. These banners are similar to requests for permission to access a person's location or send push notifications. They provide a banner that asks the user if they want to add the progressive web app to their homescreen (**FIG 4.4**).

Firefox and Samsung Internet have implemented subtle badges that let people know that the website they are currently viewing can be installed to the homescreen. These badges show up in the menu bar. In Samsung's case, the star icon that lets someone bookmark a page is replaced with a plus symbol that can be used to either bookmark the page or add the app to the homescreen. Firefox added a new icon to the menu bar that combines a house with a plus sign (**FIG 4.5**).

The add-to-homescreen badges that Firefox and Samsung use are not as obvious or intuitive as the banners in Chrome and Opera. In the case of Samsung, someone has to notice that the icon in the address bar has changed from a star to a plus symbol, and then be curious enough to tap on the button to find out what the change means. Firefox has added a bit of instruction the first time someone encounters a progressive web app to introduce them to the new button (**FIG 4.6**).

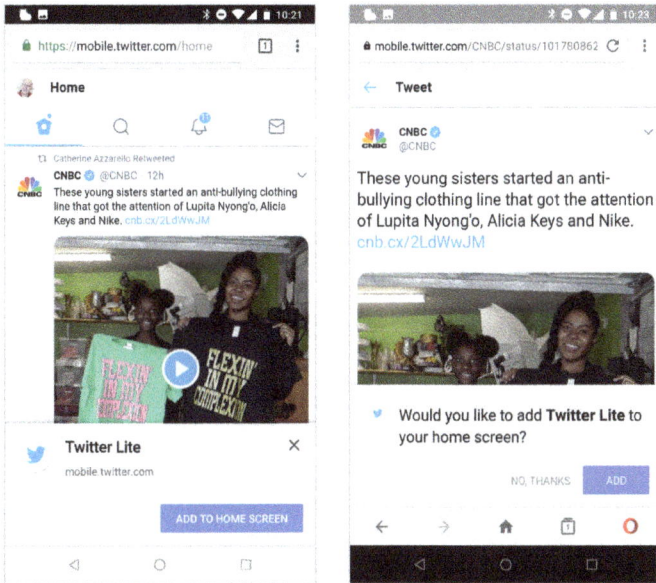

FIG 4.4: When progressive web apps first launched, Chrome (left) and Opera (right) automatically prompted users to install them using add-to-homescreen banners. This behavior has changed over time.

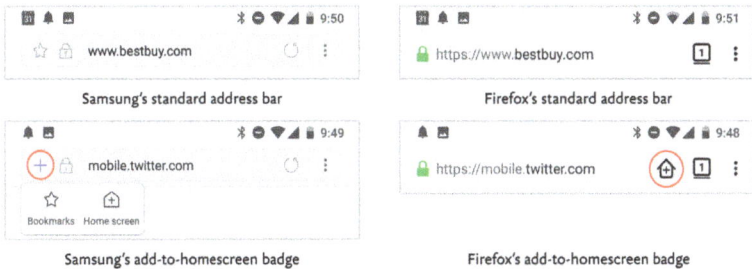

FIG 4.5: Samsung (left) and Firefox (right) use add-to-homescreen badges in the address bar to let people know that they can install a progressive web app.

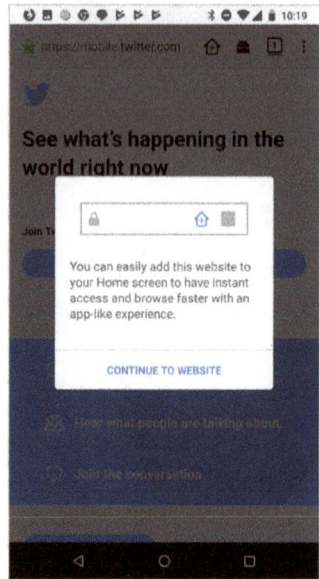

FIG 4.6: The first time someone visits any progressive web app, Firefox provides instructions about how the new badge can be used. It remains to be seen if this instruction is enough to train people to watch for the badge in the address bar.

Manifest requirements

By design, the Web App Manifest specification doesn't provide web developers with an API to install apps, nor does it explicitly define what signals browsers should pay attention to when deciding to let a user know that an app is installable. Browsers are experimenting with what works best. Chrome started with automatic app install banners, but is moving to using ambient badging combined with app install banners that only show up when the app requests them. Microsoft has announced its intention to add badging to Edge, but it's unclear what form that will take.

For badges, both Samsung and Firefox require HTTPS and a valid manifest file. Samsung explicitly requires that the manifest contain `name` or `short_name`, `start_url`, icons that are at least 144 × 144 pixels, and a display mode of standalone or fullscreen. Samsung also requires a service worker be present.

Firefox is the most forgiving of the browsers in terms of what it expects in the manifest. Firefox looks for a 192 × 192 pixel icon, but will accept smaller icons if they are available. Otherwise, so long as the manifest is in a valid format, Firefox doesn't require any additional members. Fair warning, though: if you don't supply a name or short_name, Firefox will use the URL as the name of the app. It is hard to imagine a good progressive web app experience without names and icons.

The criteria that Chrome and Opera require are similar to what Samsung uses. One small difference is that Chrome asks for icons in the form of a 192 × 192 pixel PNG with the MIME type declared using the type member. Chrome will also soon show the add-to-homescreen banner if the minimal-ui display mode is used.

Engagement thresholds

The biggest difference between the criteria all these browsers use for add-to-homescreen prompts is that Chrome and Opera require sites to pass a threshold of user engagement. Ideally, checking for user engagement will prevent users from being spammed with add-to-homescreen banners for sites they don't care about.

Originally, in both Chrome and Opera, before the add-to-homescreen banner was displayed, a user had to visit a website twice with at least five minutes between visits to demonstrate that the user was engaged with the website. Google later conducted experiments and found that they could show the add-to-homescreen banner earlier and people still installed apps at a similar conversion rate. Because of those experiments, Chrome prompts people to install progressive web apps shortly after they start interacting with a website.

Despite this, you can expect that the criteria for installability will become *more* strict as more progressive web apps are built. On his blog, Alex Russell wrote:

In general, installability criteria are tightening. Today's Good-To-Haves may become part of tomorrow's baseline. The opposite is unlikely because at least one major browser has made a strong commitment to tightening up the rules for installability. (http://bkaprt.com/pwa/04-06/)

Managing your prompts

What does this mean for your progressive web app?

First, if you don't see an add-to-homescreen banner immediately, don't assume that your progressive web app has a bug. It's possible that you haven't passed the user engagement heuristic yet. You can bypass this user engagement check by going to chrome://flags or opera://flags in Chrome or Opera respectively and enabling the "bypass user engagement checks" setting. Once that setting is enabled, you will see the add-to-homescreen banner immediately, so long as you have a valid manifest, service worker, and HTTPS.

You can also manually trigger the add-to-homescreen prompt in Chrome by tethering an Android phone to your computer and using the remote debugger. The manifest section of the Application tab in Chrome DevTools has a link to force an add-to-homescreen prompt (**FIG 4.7**).

Second, you should consider whether it makes sense to control when a visitor is presented with the add-to-homescreen prompt so that you can present it when someone is most likely to install your app.

Flipkart delays the add-to-homescreen prompt on its progressive web app until after someone completes an order. The prompt appears on the order confirmation page, in the context of helping the user keep track of their order. Asking at this critical moment leads to a threefold increase in users installing their app (**FIG 4.8**) (http://bkaprt.com/pwa/04-07/).

Intentionally choosing the right time to display the prompt, as Flipkart has done, is a recommended-but-optional practice, as of this writing. However, upcoming changes to add-to-homescreen prompts in Chrome will soon make this required.

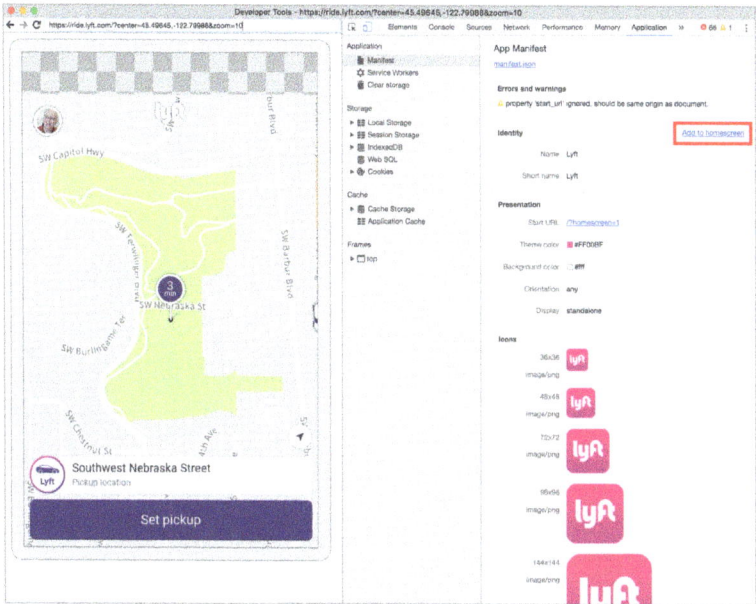

FIG 4.7: You can manually trigger an add-to-homescreen prompt on a tethered device using the Application tab in Chrome DevTools.

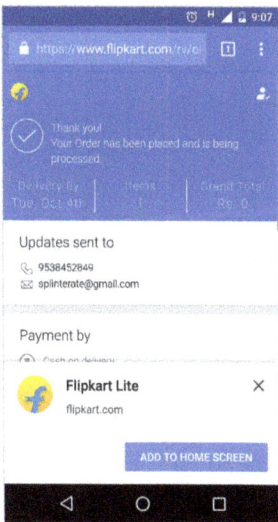

FIG 4.8: Flipkart delays the add-to-homescreen prompt until after someone places an order.

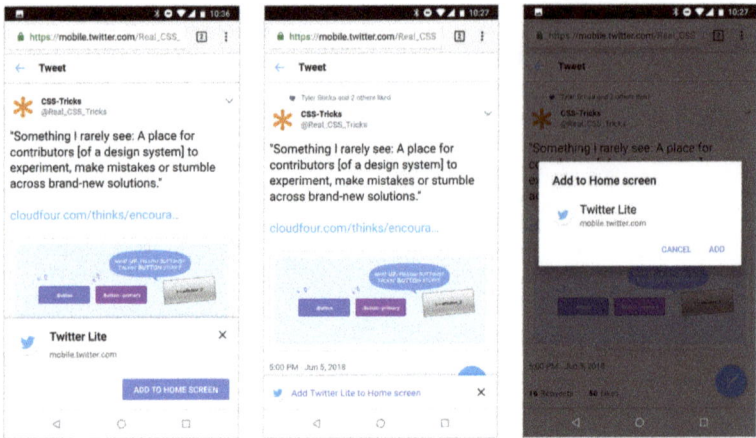

FIG 4.9: Chrome 67 automatically displayed an add-to-homescreen banner (left). As of Chrome 68, if an app meets the user engagement threshold, developers must proactively request the add-to-homescreen prompt (middle) before the modal will be shown (right).

As of Chrome 68, Chrome will no longer automatically display add-to-homescreen prompts (**FIG 4.9**). The banners will be replaced temporarily by a mini-infobar. Eventually, Chrome will add an icon to the address bar (similar to the ambient badging in Firefox and Samsung browsers) and the mini-infobar will be removed (http://bkaprt.com/pwa/04-08/).

Ambient badges are easily overlooked, so if you want to make sure people know to install your app—and you don't want to rely on ambient badging—you'll need to figure out the optimum time to ask someone to install your app.

When your site has passed all of the criteria for an add-to-homescreen prompt to show up, the browser will fire a before-installprompt event. In JavaScript, you can listen for this event and prevent the browser from displaying the prompt at an inopportune moment. You can then trigger the prompt at a time of your choosing by calling prompt() on the previously saved beforeinstallprompt event.

Both browsers and site owners need to be thoughtful about asking users to install the app when they are likely to do so. We're going to be much more successful at getting people to install our apps if we ask them in a context where they will want to say yes. Plus, as we discuss in Chapter 6, when it comes to push notifications, there are risks associated with asking for permission when you are unlikely to succeed.

APP STORES

The only way that someone can install a native app is via the app store for that platform. One of the big benefits of progressive web apps is that app stores are entirely optional—unlike with native apps, your progressive web app doesn't need to be approved by external gatekeepers to exist.

But there are benefits to having your progressive web app listed in app stores. For one thing, app stores help with discovery, providing another way for people to find your progressive web app. And while preparing your app for store listing may require a little more work, you may also gain access to fancier APIs. Not all app stores list progressive web apps as they are, though, so let's look at what it takes to break into them.

We can look to the Microsoft Store for an example of an app store that is already treating progressive web apps as equals to native apps. Hopefully, other companies will follow suit.

One of the benefits available to progressive web apps in the Microsoft Store is the full suite of Windows APIs. As long as the progressive web app was installed via the Microsoft Store, it can request permission to access the user's local calendar, contacts, and other features that aren't possible from inside a web browser yet.

Be careful though. These are not web standards. They will only work on Windows and only when the progressive web app is installed from the Microsoft Store. Install the same app directly from the browser and you won't have access to these proprietary APIs.

This doesn't mean that you shouldn't implement these APIs, but it does mean that you need to think carefully about how to do so without breaking the experience for non-Windows devices. This is an opportunity for progressive enhancement. Your progressive web app should have a baseline experience that works *everywhere*—then, if your code detects access to Windows APIs, you can enhance the experience.

Microsoft adds progressive web apps to its store in two ways. First, you can submit your progressive web app to the store in a process similar to how native apps are submitted. You'll need to generate an AppX file—the file format used in the store—and a Windows Dev Center account. To facilitate this, Microsoft has a free tool called PWA Builder that will generate an AppX and help you submit your app (http://bkaprt.com/pwa/04-09/).

Second, Microsoft may add your progressive web app to its store without you doing a thing. When Bing encounters progressive web apps on its crawls, the apps are evaluated and—if they meet a quality threshold—automatically added to the Microsoft Store. This is something entirely new to progressive web apps, and it is only possible because progressive web apps are regular websites with manifest files that search engine spiders can discover.

It's conceivable that your progressive web app could be in the Microsoft Store without you even knowing it. If that's the case, you can claim your progressive web apps in the Microsoft Store. Doing so will allow you to see analytics and other benefits associated with being in the store.

There's nothing that says that progressive web app stores need to remain the exclusive domain of operating systems vendors. Manifest files are easy to find. We will likely start seeing third-party app stores and directories that will offer additional opportunities to promote your progressive web app.

Native wrappers

The app stores for Google and Apple haven't yet opened their doors to progressive web apps the way Microsoft has—one can hope they follow suit. But while Google promotes progressive web apps, it also promotes Android apps, and thus may not want progressive web apps in the Google Play store. Apple, meanwhile, already has a reputation for closely curating its app store, and favors apps built using its native frameworks; it's likely that Apple will continue to keep progressive web apps out of the store.

So until the day arrives when you can submit your progressive web app to the Google Play store or the iOS App Store, what can you do if you want to be listed? Your best bet is to bundle your progressive web app in a native wrapper before submission.

Native wrappers have existed since the creation of PhoneGap shortly after the release of the iPhone App Store in 2008 (http://bkaprt.com/pwa/04-10/). PhoneGap and similar frameworks provide a template for using an embedded web view inside of a native application, with the assumption that the web view will do the heavy lifting of the application. It creates a wrapper for the web application that allows you to submit it to native app stores.

PhoneGap also provides plugins to enable features that are available on the native platform, but not available for the web. For example, a user's contacts are not normally accessible to a browser, but they can be via a JavaScript API with a PhoneGap plugin installed. If you don't need PhoneGap's plugins, Microsoft's PWA Builder can also generate basic native wrappers for iOS and Android.

Google recently announced Trusted Web Activity, a tool for wrapping a progressive web app and shipping it as an Android application (http://bkaprt.com/pwa/04-11/). As Jason Miller, creator of Preact and member of the Chrome team, put it, "Trusted Web Activity is basically a built-in PhoneGap for Android" (http://bkaprt.com/pwa/04-12/).

To use Trusted Web Activity, you need to declare that the app is associated with the domain and that the domain is associated with the app. This bidirectional verification prevents someone from submitting any random website as their own, which could be a security risk. Imagine for a moment an Android app that wrapped a bank site and then logged everything that someone typed on the bank site. Verification gives Google the confidence that the person who owns the app also owns the site.

When the app is trusted, cookies and storage can be shared between Chrome and the Trusted Web Activity app. This means that if a user was already logged in on the website in Chrome, they will automatically be logged in when they launch the Trusted Web Activity app. In the future, this level of trust may make Google feel comfortable exposing more sensitive APIs to verified progressive web apps that are using Trusted Web Activity.

Raising the bar

When you publish your progressive web app in a native app store, you must abide by all of the store rules, including those for content bundling and in-app sales restrictions. These restrictions can mean additional development work beyond simply wrapping your app in native code, and might raise obstacles that technology can't solve. For example, Apple only allows native apps to sell subscriptions via its own subscription tools, and those tools don't allow you to sell a subscription to a company instead of an individual. There's no code you can write that will solve a mismatch between your business model and the constraints of an app store.

Keep in mind that once you're in a native app store, user expectations will go up. If someone adds your progressive web app to their homescreen after visiting your website, they know what to expect. But if someone installs it via a native app store, they will likely expect it to look and feel more consistent with the design language of the device they're using. The bar will be higher. Make sure you're providing the best experience possible, or your app rating will suffer.

MARKETING YOUR APP

If you step back from the specific details, most of what we've been talking about in this chapter is about how you market your progressive web app. The goal of your marketing plan shouldn't simply be to get people to install your application—that's just one stage of a long-term relationship between you and your customers. Instead, your marketing plan should address each stage of the customer relationship.

The first stage is to get people to your website, typically via traditional web traffic generation activities like search engine optimization, social media engagement, and advertising. This is the first step in your funnel, and you should try to get as many people to your website as possible.

Every person who visits your website will be getting your progressive web app experience. They don't need to install your app to benefit from your work. You can offer them push notifications, offline support, and other enhancements that will entice them to continue using your website. You may have many loyal, valuable customers who never install your app on their homescreen.

But for those who might be inclined to install your progressive web app, you should set up a manifest file that reflects your brand, and consider the best time to prompt them to install your app. Look for other places to promote your application, and consider whether it makes sense to list your progressive web app in app stores.

In short, don't get hung up on people installing your progressive web app—but make sure that if they *are* interested in installing it, you've made it easy for them to do so. Either way, it's time to start using the offline capabilities of progressive web apps to make your website faster and more resilient.

5

OFFLINE

IT'S COMMON FOR PEOPLE TO LOOK AT the offline feature of progressive web apps and think that it doesn't apply to their website. Sure, it might be nice if people could access our content offline, but it isn't *necessary*.

Yes, service workers are fundamentally about caching assets, and yes, that makes them available even if someone is offline. But "being offline" is one of the least common scenarios. It is far more likely that someone will be accessing your website on an unreliable or slow network connection. Even if someone has a fast connection one moment, there is no guarantee they will continue to have one the next.

Service workers give us the ability to store things offline and take over network requests so we can route them whatever way we deem best. This allows us to build web apps that are more resilient to whatever network conditions exist. By storing the stuff needed to build the application offline and retrieving it from the local cache, we can minimize how much data we need to get from the network. If you already have everything you need to render your application locally, with the exception of the small bit of content that is unique to a given page, then even a person on a 2G connection can have a fast and reliable experience.

Storing assets offline also enables our users to continue their sessions wherever they left off. The best native apps show you what information they already have available from the last time you used the app, and then let you know that they're going to get fresh stuff on your behalf.

Service workers can do everything from improving a site's performance and reliability to making an entire application work offline. The way you use service workers on your website will affect what you communicate to your users and how complex your progressive web app will be to build. Let's look at some of your offline options.

CACHING STRATEGIES

Adding a service worker means that you are taking on the responsibility for managing the cache—including knowing when to remove files from the cache and ensuring that updated files are retrieved. It likely means adding formality and process to how you roll out updates. If you already have a regular process for shipping new versions of your code, naming conventions for those versions, and ensuring that browsers and content delivery networks flush their caches of any old code, then you're in a good position to add service workers to your mix. If not, you likely have some organizational process work to do in addition to the technical development.

But it's worthwhile work. Using a service worker to enhance performance and reliability can, and should, be done for any website. Because the work can be done behind the scenes, there is little that needs to be communicated to end users about what is being stored offline. The only thing that visitors will notice is that the website is now faster than it used to be.

At minimum, you should use service workers to cache commonly used assets, logic, and templates to increase the performance and reliability of your website. This requires thinking carefully about the documents, images, scripts, and stylesheets that are used on your site, how they change, how often, and where these assets live.

For example, let's imagine that you keep images that are common across pages in a directory called /common_images/. They may change infrequently, and when they do change, the timestamps in the filenames change, and any code pointing to the images gets updated to point to new URLs. In a situation like this, you can safely cache these images for a long time.

For other files, you might decide to retrieve them from the cache while checking the network to see if a new version exists. If a new version exists, then you might download the new version to be used the next time someone visits the page.

The Workbox JavaScript library can make setting up a service worker to do this sort of basic caching much easier (http:// bkaprt.com/pwa/05-01/). For more on the specifics of caching files and planning for fallbacks, check out Jeremy Keith's book *Going Offline*.

Recently viewed pages

The first step beyond simply caching for performance is caching recently viewed pages. The advantage of this approach is twofold. First, you don't have to try to guess which pages someone would like to access offline; you simply cache things as the person encounters them.

Second, the act of requesting the page can act as the trigger for determining what items to store in the cache. It also reduces the complexity of the service worker because you don't have to write code to precache pages and assets that the browser hasn't encountered yet.

For cloudfour.com, we chose to only cache recently viewed pages because the main reason people come to our site is to read the articles. If we tried to anticipate which articles someone would want offline access to, we'd likely guess incorrectly. If we precached top-level pages, we might force people on a metered network connection to download content they would never look at (and we know from our analytics that we do get visitors from countries where metered connections are common). By

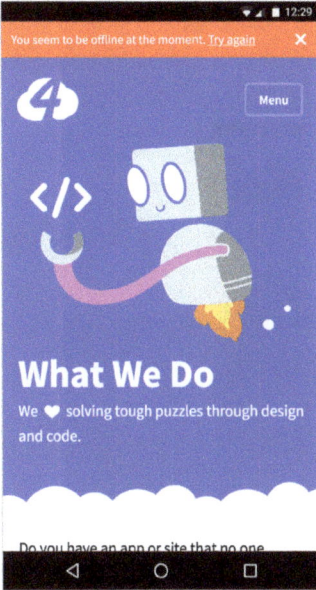

FIG 5.1: You can provide a notification when people are offline so they know what to expect.

caching recently viewed pages, readers can continue to read those pages even when they go offline.

When you start providing pages offline, you need to communicate to visitors when they are viewing an offline version of the page. On our site, we add a banner to the top of the page to communicate offline status (**FIG 5.1**).

You can also detect when connectivity returns via the navigator.onLine and backgroundSync APIs. If you detect that connectivity has returned, you can try connecting to the server, and offer your user the option to refresh the page to get the latest information.

If the content is time-sensitive, it may be important to convey how long ago the content was updated. For a blog post or article, it may not matter how long ago it was stored offline. But a currency exchange application would likely want to alert users about how long it has been since the app had been able to connect to the internet and download new exchange rate information (**FIG 5.2**).

Material Money, a demo currency exchange calculator, provides information on the last time the rates were updated (http://bkaprt.com/pwa/05-02/).

Typically, we think of a page's availability as binary, either offline or on. But for pages that pull content in via AJAX, it is possible that portions of the page may have loaded before network connectivity was lost.

Trivago handles this scenario well by showing whatever information it has available when you search for a hotel. Each hotel result has four tabs (for Photos, Info, Reviews, and Deals). If Trivago has downloaded information for a tab before you lose connectivity, you can open and close that tab without worrying about losing the information—it is stored offline. However, if you visit a tab where the content hasn't been downloaded yet, Trivago will display an offline message in the tab where the content would normally be displayed (FIG 5.3).

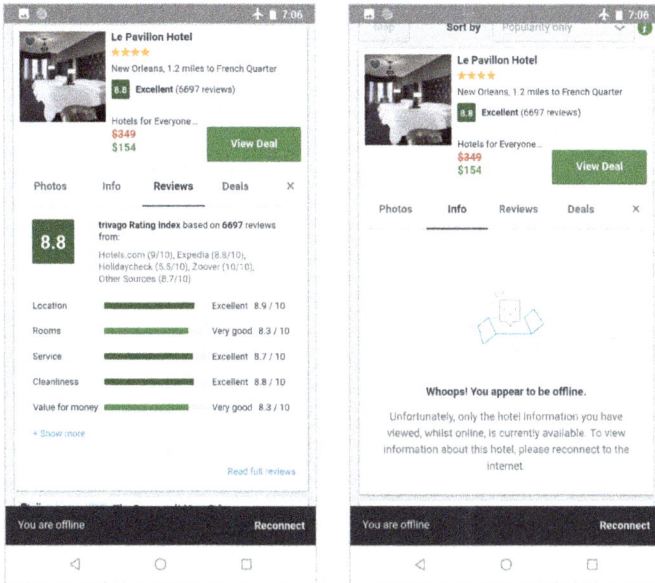

FIG 5.3: Trivago treats portions of the page as discrete areas that may or may not be available offline (left). If an area isn't available offline, a message appears in the body of the page (right), letting the user know that the area isn't available.

User's choice

Instead of trying to guess what people will want to access offline, we can ask them. News sites have started to experiment with providing users with the option to select articles that they want to read offline (**FIG 5.4**).

In addition to the toggle on the article page itself, you'll also need to provide some place where users can see what articles have been stored offline and a mechanism to remove those articles. Both storing articles offline and removing them from the cache should be accompanied by confirmation notices (**FIG 5.5**).

FIG 5.4: The *Financial Times* allows readers to bookmark articles to their myFT account for offline reading.

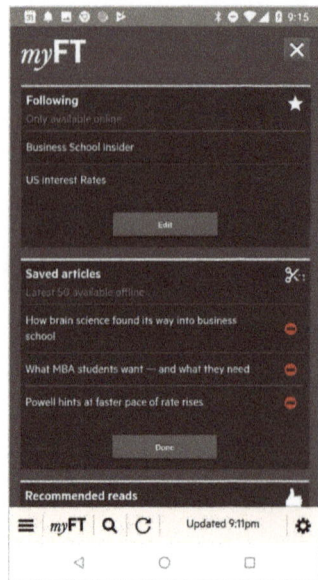

FIG 5.5: *Financial Times* readers can see the articles they've saved for offline reading and delete articles they no longer need.

Precaching

If you have reasonable confidence about what content people are likely to want offline, you can preemptively cache the necessary assets—but you must be careful. Many people access the web on slow or metered connections. Forcing someone to download a large amount of data behind the scenes the first time they visit your site is irresponsible.

How can you determine if it makes sense to precache assets? How do you build confidence about what visitors might want in the future? Your analytics data may be able to tell you where you have high-traffic page funnels. For instance, if 80 percent of visitors to a particular page visit another page afterward, you might feel confident enough to precache the content of those pages.

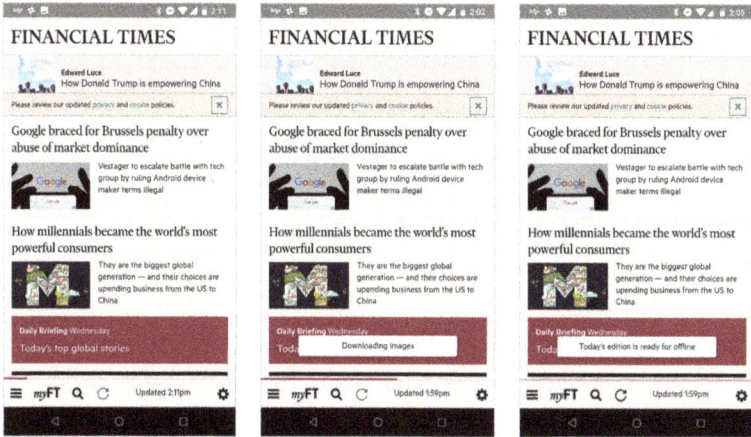

FIG 5.6: The *Financial Times*'s progressive web app displays a brief notification to let the user know that the articles are ready for offline use.

You should also keep in mind the size of the assets you want to cache. For example, a news site might precache the most important stories, but not the images, so that critical information, at least, would be available offline.

No matter how confident you are about what the typical user does on your website, not every user is going to fit the typical journey. Therefore, your best bet is to give your users visibility and control over what is cached. The *Financial Times*, for example, does a good job of informing users by showing a progress bar as content downloads, then letting users know the articles can be accessed offline (**FIG 5.6**).

This transparency lets users gauge how much time and space is being spent on images. They can also modify their settings to turn off image downloading, or automatic downloading of content altogether (**FIG 5.7**).

You may also want to notify users when there is a new version of your app available and encourage them to refresh the application. When an application has been cached, people will typically see the old version of the application for one session whenever a new version is released. For the *Financial Times*, getting people to refresh the app isn't always a pressing need,

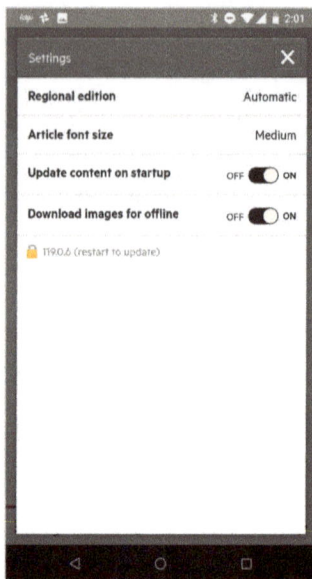

FIG 5.7: The *Financial Times* gives users control of what content should be downloaded automatically. It also lets the user know when to update to a new version of the app.

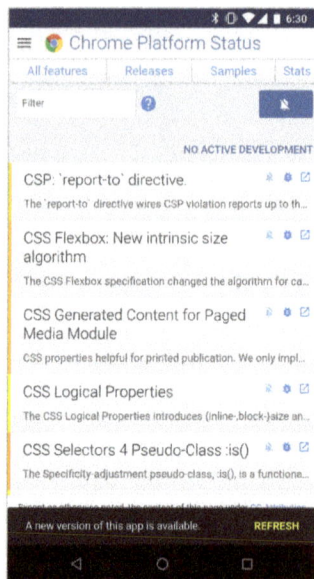

FIG 5.8: The Chrome Platform Status website doubles as a progressive web app. When new versions of the progressive web app have been downloaded in the background by the service worker, the app notifies users that they should refresh the app (http://bkaprt.com/pwa/05-03/).

so they display the version number and the message "restart to update" in the Settings area, where few users are likely to see it. By contrast, the Chrome Platform Status website displays a highly visible version notification, along with an immediate Refresh button (**FIG 5.8**).

Be honest with yourself about the value of precaching for your users. Just because we *can* precache several megabytes of an application doesn't mean we *should*. Be transparent about your caching practices, and let your users control them.

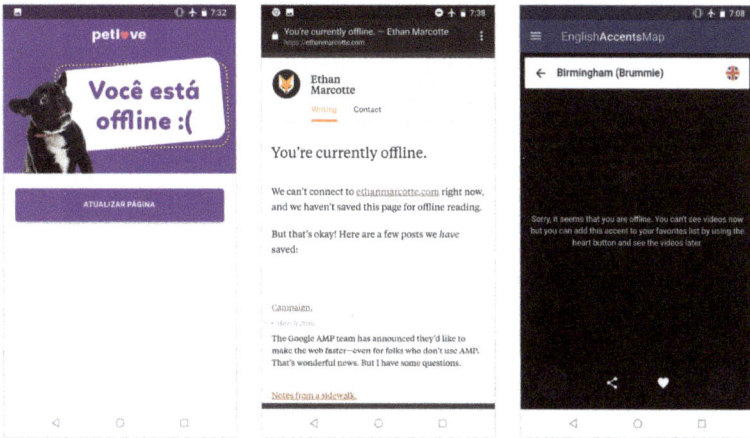

FIG 5.9: Offline fallback pages can be as simple as Petlove's (left), or add some limited functionality, like Ethan Marcotte's list of cached recent articles (middle). Similarly, English Accents Map (right) prompts users to mark videos they want to watch when they are back online.

OFFLINE INTERACTIVITY

So far we've cached some assets and let people read some things offline, but what can someone *do* offline? There are a wide range of options for the types of interactivity that you can provide users when they are offline. The starting point should be to provide an offline fallback page—a page displayed for a URL that isn't available offline (**FIG 5.9**).

The fallback page can be as simple as a message notifying the visitor that they are offline. Some sites use this as an opportunity to give users something fun to do—perhaps in the hopes that entertaining a person offline will persuade them to return when online (**FIG 5.10**).

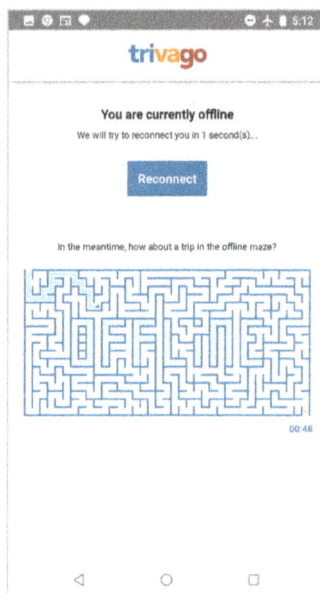

Saving offline changes

If you add interactivity to your offline pages, consider how you'll communicate to visitors what they can do while offline. For example, if your website helps people create content or edit documents, it's important to communicate what happens to the user's work when they lose their connection.

Ideally, you'd store all of the edits or changes that they make until they are back online. Doing so means adding some sort of client-side data store, likely either IndexedDB or Web Storage, in addition to the service worker. All requests to save things to your server would need to be intercepted and monitored to see if the network is available or the connection to the server has failed. Then, instead of saving it to the server, you would reroute that change to the local data store with the intention of saving it to the server at the next possible moment.

Doing this work has become easier in recent years. The Polymer team has released a web component for syncing an

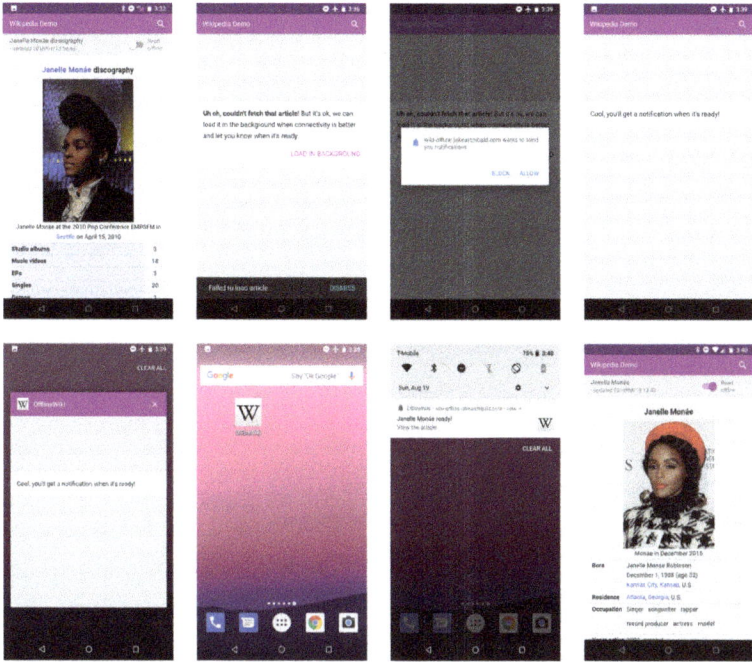

FIG 5.11: Jake Archibald's Offline Wikipedia app uses background sync to download articles when the network connection is too slow. Download requests persist even after the app has been closed (lower left). The user is notified that the content is ready via a push notification. (http://bkaprt.com/pwa/05-06/)

in-browser database with a remote database (http://bkaprt.com/pwa/05-05/). And yet, even with libraries like the Polymer component available, adding offline syncing to an existing application could be a substantial task. It might not even be feasible if your app doesn't have public APIs to sync with.

You should still notify the user that they are offline so they know the changes won't be reflected immediately. An offline notification may also give someone pause if they were about to do something mission-critical and can't risk being offline while making changes. There are few things more frustrating than losing all your work because you're no longer online.

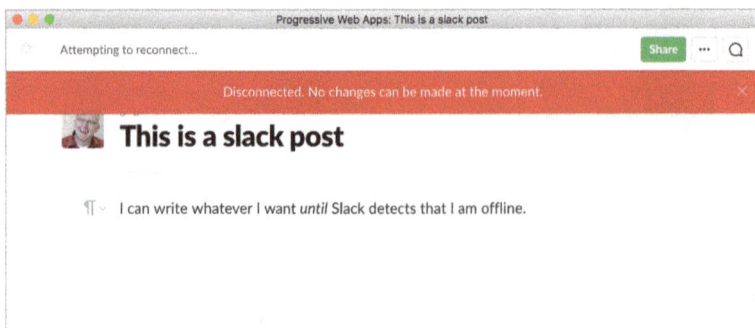

FIG 5.12: While it isn't a progressive web app, Slack provides a good example of saving people from making edits that cannot be saved while they are offline or on an unreliable network connection.

Similarly, when the network improves, you should inform the user of the status of their edits. It may be sufficient to simply remove the warning message. Or you may want to provide confirmation that the edits have been saved.

If someone closes the app before they are online again, the Web Background Synchronization specification, an extension to service workers, provides the option to sync the data in the background when a network connection is reestablished (**FIG 5.11**) (http://bkaprt.com/pwa/05-07/). When you send an email in a native app, you don't have to wait around until the email is handed off to the server—you can close the app and do something else. If you're offline, the app will continue trying until you're back online. Background sync is only available in Chrome, Samsung, Opera, and UC Browser at the moment, but is in development for Edge and Firefox (http://bkaprt.com/pwa/05-08/).

If storing changes offline and sending them to the server later is too large of an undertaking, a short-term fix can be to do what Slack's desktop app does: disable editing when the network connection is down or unreliable (**FIG 5.12**).

While this isn't an ideal solution, it does prevent someone from investing a lot of time in edits or changes that the app won't be able to save due to network conditions.

ONE SITE, MANY APPS

As you consider different options for how you use service worker technology, you may find that some areas of your website need more complex offline capabilities, while others may only need baseline performance improvement. Many websites may have content areas—such as those describing the company, history, or policies—that are important, but not critical for offline access.

One of the nice things about progressive web apps is that you can treat different areas of your website differently. A company selling building supplies might choose to speed up the performance of its website's About Us section, cache recently viewed product pages, and make its special building materials calculator available for contractors to use offline.

In fact, there may not be a one-to-one relationship between your website and the progressive web app that you want to build. How companies have handled native mobile apps is instructive here—many companies with a single application on desktop have multiple apps on mobile devices. In a desktop browser, Facebook is a single app with sections for Messenger, Events, Pages, and Groups. On mobile, they are all separate apps (FIG 5.13).

Yahoo also has multiple native apps. Many of these native apps map to Yahoo subdomains and sections of its desktop site. For example, sports.yahoo.com and finance.yahoo.com each has its own native app (FIG 5.14). Yahoo likely has different teams working on each of these subdomains. It makes sense that each is its own native app, and the same logic would apply to any progressive web apps created in the future.

Perhaps your site has one particular area of functionality that makes sense as an app, such as the building materials calculator from our fictional building supply company—it would make sense for that aspect, rather than the whole site, to live as a separate icon on a contractor's homescreen. I refer to apps like these as *tearaway apps*. They remind me of posters cut into strips at the bottom so that interested parties can tear away information and take it with them.

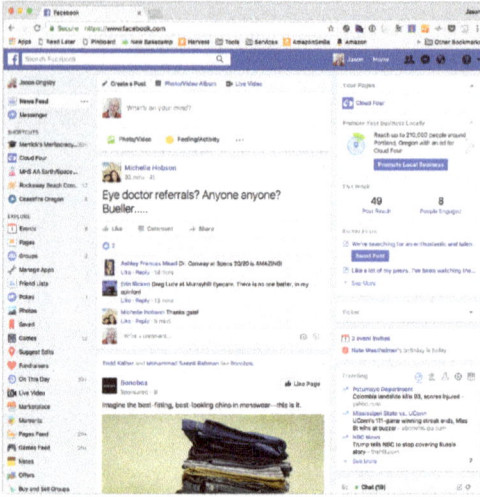

FIG 5.13: Facebook has a single desktop app (left), but several native mobile apps (right). If you have a large site, will a single progressive web app do?

FIG 5.14: Yahoo's website has many separate subdomains (right)—for example, news.yahoo.com— which map directly to separate native apps (left).

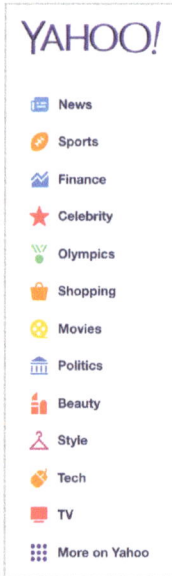

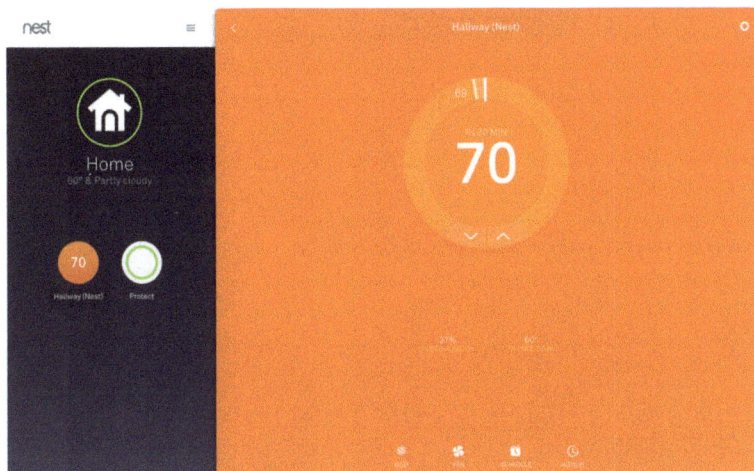

FIG 5.15: Nest's website has many areas of content, but for Nest owners, the most important are the tools that control your Nest devices.

Progressive web app technology supports the idea of tear-away apps. Individual sections of a website can have their own manifest files and service workers. With a little planning, you can treat different areas of your website as separate progressive web apps.

Nest's website is a perfect candidate for developing a tear-away app. While the website contains sections ranging from marketing material to support documentation, the most likely candidate for a progressive web app is the part of the site you see when you log in: your control dashboard (FIG 5.15).

In fact, if you log into nest.com on a mobile device, the website is nearly indistinguishable from Nest's native apps (FIG 5.16).

Visitors to nest.com would benefit from service workers being applied to every page on the site for performance reasons, but the part of the site that they would most likely want to tear away and take with them would be the part that lets them manage their Nest products.

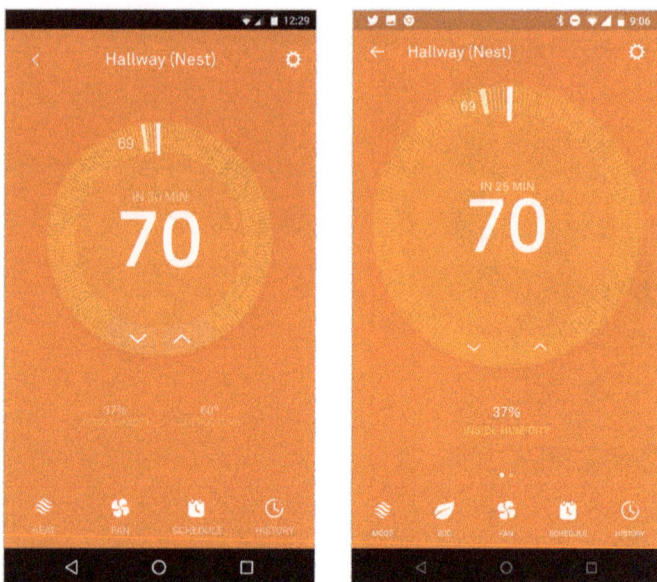

FIG 5.16: Nest's native app (left) and its responsive website (right) are nearly indistinguishable.

As you think about the areas of your website, consider whether your entire site would be useful as a progressive web app, or whether different areas of the site should be handled differently. Are there discrete areas that should function independently? There's nothing that says you can't have more than one progressive web app for your website.

From a performance perspective, every page of your site can benefit from offline caching. Past that, you'll have to determine what would most benefit your users to have available offline, and what offline strategy makes the most sense.

OFFLINE FIRST

Many of the considerations we've been talking about can be baked right into your design—if you *begin* with the assumption that your application will work offline. This is a big shift in the way we think about the web. Since the earliest days of the web, we've built and designed our applications with an assumption of network connectivity. Since moving past the dial-up era, our computers have been almost constantly online.

The growth of mobile phones, with their intermittent connections, has taught us that our network connections aren't always reliable. But while we've adjusted to this as phone users, we haven't yet learned to think that way when building web applications.

It can be difficult to graft the ability to make offline changes onto an existing application. But if you're building a new web app or have a chance to rethink an existing app, you should provide an offline experience by default. If you treat the presence of a network as an enhancement to your application, then your application will work under any condition.

When our apps are able to complete changes while offline and sync when an application is closed, it becomes imperative to communicate with users about what the apps are doing on their behalf when they don't have the app open. This is where push notifications can help.

6 PUSH NOTIFICATIONS

A DECADE AGO, MOBILE MARKETING presentations and articles prophesied that people would receive special offers on their phones as they walked past stores. The store in question was almost always a Starbucks, and the scenario was always presented as a benefit to consumers who were simply dying to receive more ads.

In truth, no one wanted that future. Perhaps enough people had seen the mall advertising scene in the movie *Minority Report* to know that having stores bark at us as we walk by isn't some glorious future, but the dystopian nightmare of capitalism run amok. We don't need additional interruptions in our lives; I doubt the marketers themselves would sign up for a barrage of push notifications as they walk past businesses they have no interest in patronizing.

While this particular fantasy faded when beacons flopped, push notifications continue to hold allure for marketers—and it's understandable why. You can connect to people anywhere in the world on the most personal device they own. You can entice them to return to your app with offers, breaking news, or other tantalizing bait.

But too many websites are rushing to send push notifications without giving users a good reason to allow them. Push notifications *can* be beneficial to both users and businesses—but only if they're done well.

While push notifications are new to the web, they're not new to users. People already have opinions about push notifications from native apps—namely, they don't like them. A survey by Localytics found that 52 percent of users think push notifications are "an annoying distraction" (http://bkaprt.com/pwa/06-01/). Push notifications have been abused by far too many companies, bombarding users with notifications they don't want, instead of thoughtful, timely information.

Implementing push notifications well requires careful consideration and planning. Sure, you can integrate a push notification service into your website fairly quickly. Like adding an email newsletter sign-up form to your website, the technical implementation usually isn't complicated; writing, editing, and maintaining the newsletter *is*.

Push notifications are no different. The technology itself isn't much use without a plan for using it.

If you want to use push notifications, but don't yet have a program in place to deliver valuable and relevant information to users, then you probably shouldn't implement push notifications yet. Let's look at some factors to consider in your planning.

PLANNING FOR PUSH

There's a reason organizations often plan push notifications for the later phases of their progressive web apps. These organizations see the value, but when they start thinking about *how* they'd like to use them, they often realize there's a much larger scope of work involved to make them useful. After all, push notifications don't write themselves.

Consider how, when, and why your customers interact with your company, and you'll likely come up with scenarios that are more complex than sending ad hoc messages. If you're an ecommerce company, perhaps you want to send notifications to customers when their order has shipped, or if they've left something in their shopping bag but didn't check out. A news organization may want to allow readers to select the topics they are interested in receiving alerts about. A sports website might want to send game-time updates from teams that the user is tracking.

What makes these scenarios useful to users is that the push notifications are personalized for their interests and triggered by external events.

While it is tempting to add push notifications to a website just because we can, the real benefits come from taking the time to figure out what will provide a tailored experience for your users. Localytics found that "users are 3x more likely to complete a conversion event if the message incorporates some kind of personalization" (http://bkaprt.com/pwa/06-02/).

In most of these scenarios, the heavy lifting happens on the server, not in the browser. The ecommerce company's order processing system has to trigger a push notification when an order has shipped. The news site's content management system needs to check new articles as they are published against the hundreds of reader profiles to see if anyone should be notified. The sports website has to monitor scores and match them to users' notification criteria. To see how complex this process can get, look at the flowchart Slack uses to decide whether or not to send a push notification (FIG 6.1).

And there are likely additional hidden requirements for your push notifications based on how you answer these questions:

- Do you want to track who receives notifications and what percentage of people respond to the notifications?
- Do you want to be able to tie notifications to authenticated users?
- Do you need a profile of users that tie their notification behavior with other touchpoints, like email?
- Do you want to be able to send marketing messages to audience segments? If so, how do you give people control so they can opt out of marketing messages they don't want?

You may not need all of those features, but once you start digging into the requirements of a notification program, you'll

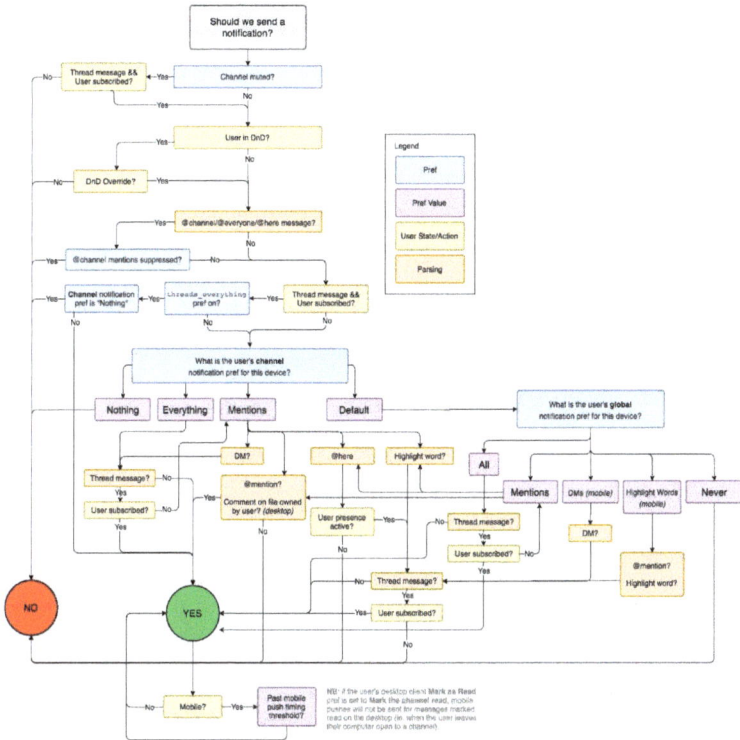

FIG 6.1: Slack's flow chart for deciding whether or not to send a push notification illustrates how the heavy lifting for push notifications happens on the backend, not the front end (http://bkaprt.com/pwa/06-03/).

likely find that there is more work to be done on the backend systems than it seems at first.

IMPLEMENTING PUSH

Because progressive web apps are multi-device and multi-browser, deciding where to send a notification can be as complicated as deciding whether or not to send one in the first place. To send push notifications, you have to connect to each

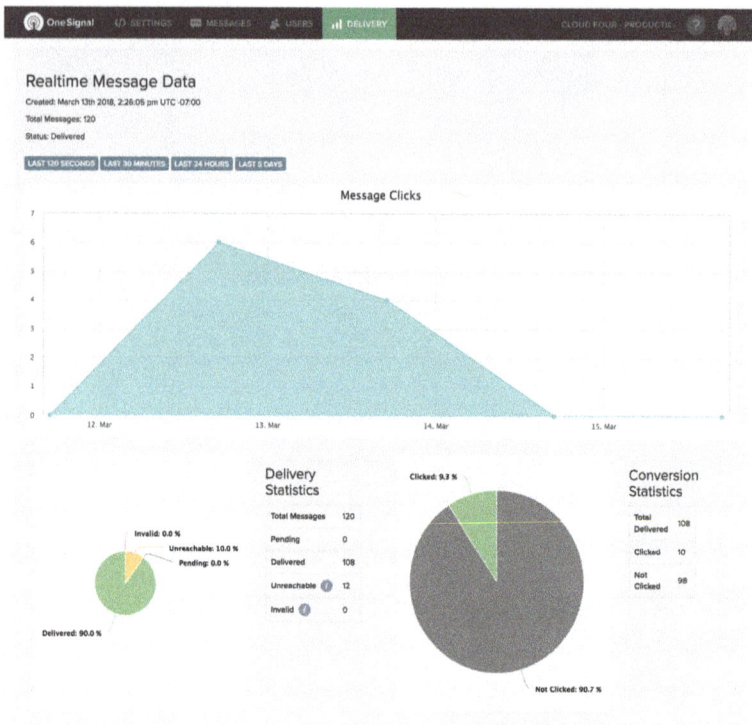

FIG 6.2: Most push notification services provide analytics and audience segmentation tools, like these from OneSignal.

browser's separate push notification end points. For instance, if you want to send a push notification to a Firefox user, you have to bundle the message you want to send with the subscription object for that user and hand those items to Firefox at the specific URL Firefox provides.

To simplify portions of the push process, you can sign up for a push service provider. These service providers shorten the implementation time by connecting to all of the service end points on your behalf. Many offer built-in analytics and marketing tools (FIG 6.2). For all but the largest of companies, using a push service provider will likely make sense. Do your research to make sure these companies support the full set of conditions and features your app requires.

But even if you use a push service provider, you'll still need to have a plan for using push notifications and a team to manage them. And if you want to integrate push into existing functionality and business processes, you'll need someone to write code that watches for a change in order status and then connects to the push service provider to notify your customer.

By comparison to implementing those backend systems, what happens in the browser is small potatoes. Front-end developers have it easy: in JavaScript, we can test to see if push notifications are available in the current browser. We then request permission to send push notifications, and, if the user approves, the browser will give us a unique subscription object. We then store that object so we can make sure the right messages get to the right person.

GIVING USERS CONTROL

You may have determined how and why to send push notifications, but you also need to give users a reason to *trust* you to send those notifications responsibly. Too many websites ask for permission to send push notifications the very first time someone visits the site (**FIG 6.3**). This is like asking someone to marry you on the first date—shouldn't we get to know each better first?

Until recently, there wasn't any disincentive for asking for push notification permission on the first visit. Sure, a user might block the permissions request, but few do: Google's internal metrics show that 90 percent of people don't choose to allow or block permission requests—they simply ignore or dismiss the permission modals (**FIG 6.4**) (http://bkaprt.com/pwa/06-04/).

This is understandable. People want to get to the content. They don't want to stop to figure out what is being asked of them. So if they have the option, they'll skip making a choice—which means the site will, annoyingly, keep asking them for permission on every visit.

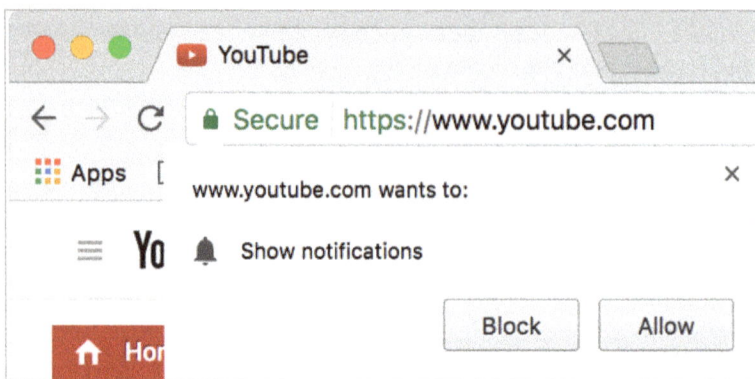

FIG 6.3: Even major websites like YouTube have been known to ask for permission to send push notifications on someone's first visit.

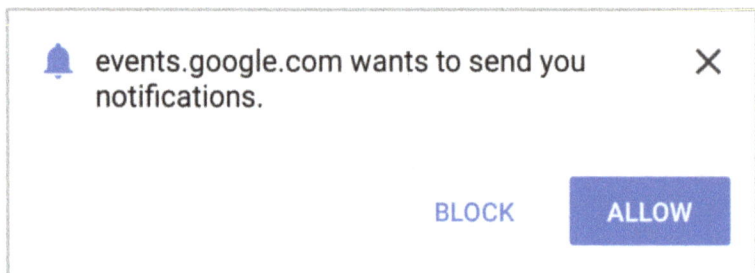

FIG 6.4: Chrome's previous permissions modal allowed users to dismiss the modal without making a choice by tapping on the close button or by tapping on another area of the screen.

To combat this, Chrome changed the way it asks people for permission as of Chrome 63. The modal now takes up the entire screen and can no longer be ignored or dismissed (FIG 6.5).

While this seems like a small change, removing the ability for people to ignore or dismiss the permissions prompt will force websites to change because there is a steep price to pay if someone blocks a permission request.

Once someone blocks your permission request, you don't get to ask again. That's it. Game over. You lose.

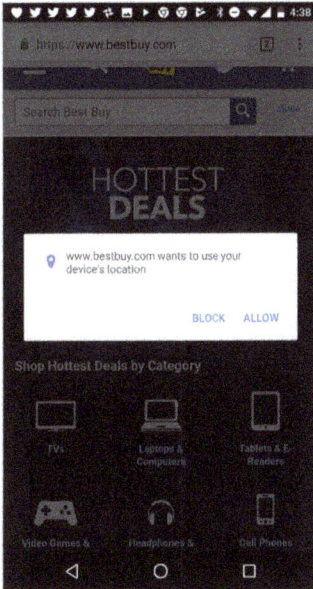

FIG 6.5: Chrome's new permission modal forces users to choose to allow or block the requested permission. There is no option to dismiss the modal.

This makes it incredibly important to ask for permission when you have a reasonable level of confidence that someone is likely to say yes. Where you find that confidence will vary. The amount of time someone spends on a site, a repeat visit, viewing a particular page, completing a transaction—any of those could indicate that someone is more likely to be interested in push notifications.

The way Twitter asks for permission to send push notifications is a textbook example of doing this well. First, they only ask about push notifications when the user visits the notifications tab. Then, instead of immediately asking the browser to provide the permissions prompt, Twitter uses an overlay to tell the user *why* they want to send push notifications. Finally, only if someone taps the button to turn on push notifications does Twitter finally execute the code to request notifications permission from the browser (FIG 6.6).

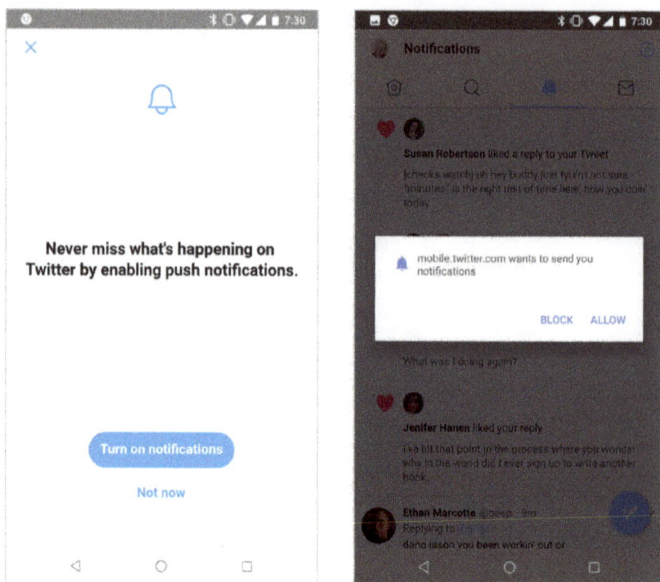

FIG 6.6: Twitter only asks the browser for permission to send notifications (right) after the user has indicated interest by visiting the notifications tab and asking to turn on notifications (left).

From Twitter's example, we can take away four key steps for respectfully—and effectively—asking the user for permission to send push notifications:

1. Ask when you can be reasonably certain the user will say yes.
2. Provide information on what the push notifications will do.
3. Prompt the user for an action to indicate they want push notifications.
4. Only ask the browser to show the permissions prompt modal if the user has indicated their interest.

Doing these four things will increase the odds that users will sign up for your push notifications, and greatly decrease the chance your site will be blocked.

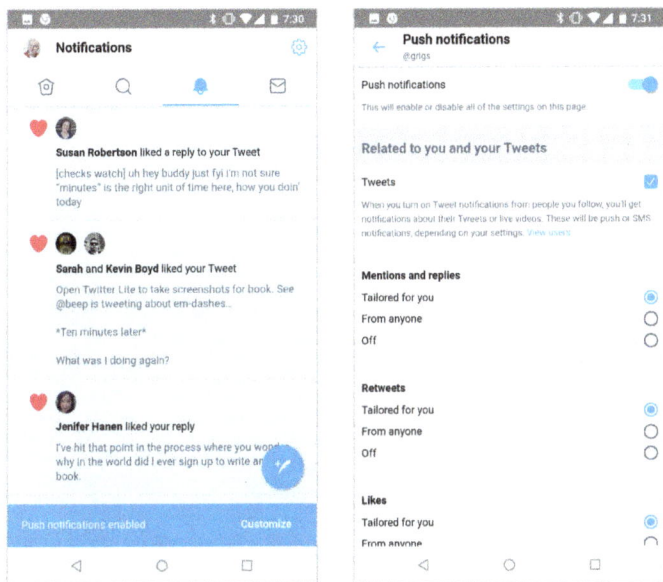

FIG 6.7: Twitter provides users with a link to modify which notifications they want to see (left). This increases the chances that someone will fine-tune their settings (right), instead of blocking Twitter's notifications.

Once a user has given you permission to send them notifications, though, your work isn't done: you also need to give them the ability to control their notification settings.

You should make it clear to users how they can modify, or even turn off, their push notifications. That's not just respectful of your users, but also smart for your business: if you don't provide people with these tools, they may use the browser's settings to block push notifications from your site—and then you'll be locked out of communicating via push from that point forward.

Here again, Twitter provides a useful example. Immediately after signing up for notifications, users are notified that the sign-up was successful, and are linked to a settings page where they can control what kinds of notifications they want to receive (**FIG 6.7**).

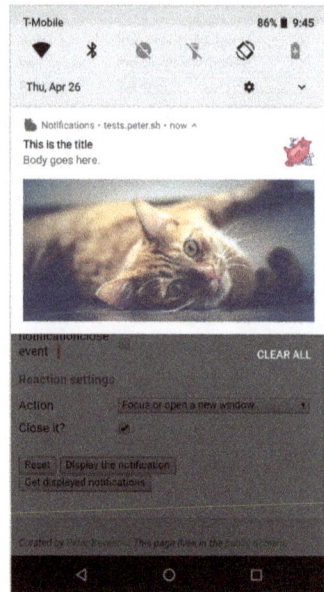

FIG 6.8: Push notifications can be customized with icons, badges that show up in the status bar, and photographs. Most companies aren't taking full advantage of the design opportunities that push notifications offer.

DESIGNING PUSH NOTIFICATIONS

We tend to think of push notifications as simple bits of text—in other words, not much to design. However, that's not entirely true. Depending on what browser and platform you're using, push notifications can be much more engaging than a few words on a white box.

Push notifications have several visual options:

- You can set a `title` and `body` for the notification. The `title` will often be displayed in a larger and bolder font.
- You can declare if the text should be displayed left-to-right or right-to-left.
- An icon can be specified for inclusion in the push notification itself, and, on Android, a badge that appears in the status bar (**FIG 6.8**).

FIG 6.9: Brazilian company Petlove's push notifications stand out thanks to their images and action buttons.

- You can also set a picture for the push notification. Petlove does this for its push notifications, which helps them stand out among other notifications (**FIG 6.9**). Keep in mind that not all platforms will display the image with the same crop, or even at all.

And design isn't purely visual: you can also specify a vibration or sound that you would like to play when the notification is received—though I've never seen anyone actually capitalize on that ability.

You can also specify actions that the user can take in response to your notification. The actions are displayed as buttons with the title and icon that you provide (**FIG 6.10**). Browsers limit the number of buttons you can display to fit with native UI, so you should check Notification.maxActions to find out what the limit is.

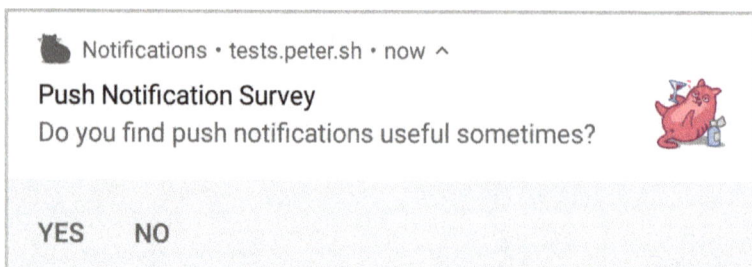

FIG 6.10: Push notifications can also present options and perform different actions based on the user's choice.

Unfortunately, not all of these features are supported in every browser. Some browsers even behave differently depending on the platform they are on. Thankfully, we can use feature detection in JavaScript to test what options are available. For example, we might check to see if a browser supports actions and, if it does, display two buttons; for browsers that don't support actions, we could instead take the user to a web page that provides the two options in the body of the page.

To get a sense of all of the features available to you when designing your push notifications, try the Notification Generator progressive web app by Peter Beverloo, one of the editors of the Push API specification (**FIG 6.11**) (http://bkaprt.com/pwa/06-05/). This app lets you turn different notification options off and on to see what they look like and test their availability.

Because web push notifications are a new area, there are many inconsistencies between browser implementations and the recommendations. Matt Gaunt, a Google engineer, has begun documenting many of them in the *Web Push Book*, which you can read online (http://bkaprt.com/pwa/06-06/). It's telling how young notifications are when Matt writes sentences like "sadly there aren't any solid guidelines for what size image to use for an icon," and "given the differences in ratio between desktop and mobile it's extremely hard to suggest guidelines" (http://bkaprt.com/pwa/06-07/). When I read statements like this, I interpret it as a caution to make sure we test thoroughly on multiple devices and platforms.

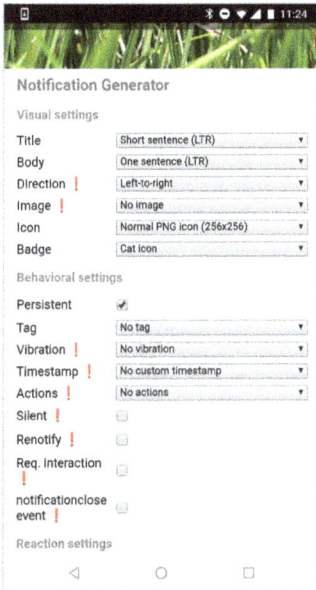

FIG 6.11: The Notification Generator progressive web app from Peter Beverloo is a wonderful tool for testing out all of the various push notification options.

Don't get discouraged by the inconsistencies. Even if a browser supports the most basic push notifications, it can still be worth your time to implement them as a way to bring people back into your progressive web app. As we discussed in Chapter 2, companies that have implemented push notifications have seen increased sales and engagement from people who choose to receive them.

USING PUSH RESPONSIBLY

Browser makers have some reservations about giving push notification power to all websites. Understandably, they fear that users—who are already annoyed with push notifications—will be frustrated that seemingly every website they visit now wants them to sign up for more—and even worse, most websites won't bother to establish a relationship with them before asking.

As of Firefox 59, Mozilla has added a feature that allows users to block any permission requests for notifications or access to microphone, camera, or location. This feature, along with the changes to the way Chrome requests permission, is an acknowledgement that users are tired of websites needlessly asking them for intrusive information.

Some browser engineers have started looking at ways to track push notification acceptance rates and other factors that might indicate whether or not a website is using push notifications responsibly. In the future, websites with a poor push notification reputation score may be blocked from even asking for permission in the first place.

Watching these changes depresses me. The web finally gets access to the sorts of capabilities that we've been clamoring for, and web developers immediately begin abusing them, forcing browser makers to lock them down and take action to protect users. I don't blame the browser makers; I expect better from developers.

So be better. Don't just staple on a third-party push notification library, ask for permission on the first visit, and spam with constant messages that users can't shut off. You're going to damage your business, irritate your customers, and endanger everyone's ability to use push notifications for legitimate purposes.

Give users control of their experience. Ask them for permission at the right time. Know how, why, and when to send notifications. By being respectful and adding value, push notifications can be an effective part of the web experience for everyone involved.

BEYOND PWAS

AS WE'VE SEEN, PROGRESSIVE WEB APPS have an expansive definition—or definitions, depending on how you look at it. Because of this, there are many new web technologies that are often discussed in the same breath as progressive web apps, despite not actually being part of them.

Whether they fit into the definition or not, there is a good chance that someone will suggest these technologies as you explore your progressive web app because they have similar goals to PWAs: they focus on providing a faster and smoother user experience.

Let's take a closer look at three of these technologies—AMP, the Credential Management API, and the Payment Request API—and their relative merits.

AMP

The Accelerated Mobile Pages (AMP) Project is a Google initiative to promote faster pages (http://bkaprt.com/pwa/07-01/). It uses a custom HTML format to limit what can be included in a web page to ensure that the page will load and render quickly.

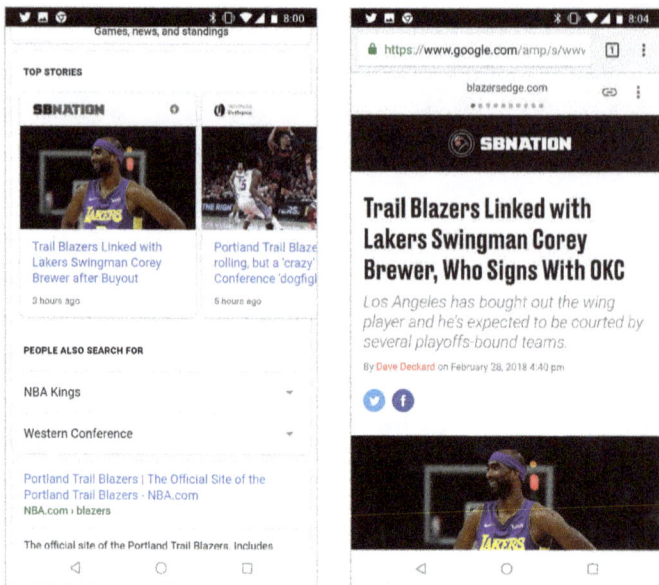

FIG 7.1: The Top Stories section on Google search results provides cards for AMP stories (left). The stories are then displayed in Google's AMP reader carousel (right).

You've likely seen AMP pages in your Google search results in the past—AMP news stories appear before regular search results, and feature a small lightning bolt badge (**FIG 7.1**). And it isn't just for news sites: Google says that "over 60% of all the clicks to AMP pages in Google search go to non-news pages" (http://bkaprt.com/pwa/07-02/). Apps like Twitter and Facebook are also beginning to use AMP pages as faster alternatives on mobile devices.

At the 2018 AMP Conference, Google announced that there are over 5 billion AMP pages (http://bkaprt.com/pwa/07-02/). Google points to that number as evidence of the success of AMP. Certainly it's hard to argue with billions of pages that load faster on mobile devices.

Challenges with AMP

But AMP is also a problematic technology, negatively impacting the connectivity and usability of the web. While the goal to create a fast web experience is admirable, AMP effectively creates two tiers of search results—and the only way to participate in that higher placement is to build AMP pages. Google's monopoly position in search all but forces organizations to conform to its standards.

The implementation of AMP pages within Google search results also presents navigation problems. The AMP carousel doesn't display content hosted on the original host domains; instead, the content is served up from Google's content delivery network on a Google domain. If you try to copy the URL from the address bar, you're sharing Google's URL, not the original source document. Additionally, AMP pages maintain a standard design across all pages, regardless of origin or brand, making it difficult to distinguish between reputable and non-reputable news sources (http://bkaprt.com/pwa/07-03/).

Carousel implementation is buggy as well, especially in Safari on iOS. The AMP carousel breaks the ability to tap the status bar to return to the top of the page, and renders the Find on Page option unusable. The carousel is often not recognized as a separate page; swiping to return to the previous page of search results will return users to the page before the search results, forcing users to enter their search terms again.

Growing concern from the web community about AMP and Google's promotion of it led to an open letter to Google signed by hundreds of web developers, including me (http://bkaprt.com/pwa/07-04/). In response, in March 2018, Google announced that it would attempt to standardize the lessons from AMP so that it could open up the search carousel and other features to non-AMP content.

It remains to be seen whether Google can fulfill its commitment to provide ways for non-AMP content to be treated the same as AMP content. Google will need to convince other browser makers to participate in the creation and definition of standards for these non-AMP pages. Even after this recent announcement, a lot of skepticism persists among web developers.

Working with AMP

Given all the problems with AMP, why would anyone use it? That's one of the thorniest issues. Google holds enough influence over inbound traffic that if it says *jump*, many organizations must ask *how high*. Chartbeat found that "Mobile Google Search traffic to our AMP-enabled publishers is up 100% over the same time frame, traffic to publishers not using AMP is flat" (http://bkaprt.com/pwa/07-05/). No matter how much organizations object to AMP, it may be hard to say no.

If your organization is going to support AMP, use it to give your progressive web app a head start. There are two ways that AMP can be integrated with your progressive web app.

The first is to use an AMP element called `amp-install-serviceworker` to install your progressive web app's service worker in the background while someone is looking at your AMP page. Your service worker can then download assets that will be needed for your progressive web app so that when someone follows a link from the AMP page to your main site, your progressive web app is ready to go. This will work even if the AMP page is viewed from Google's cache in the AMP carousel.

The second option is an idea promoted by the AMP team to use AMP documents as the content container in your app shell. In this scenario, the progressive web app almost acts like an AMP viewer. The same AMP document can be used for search results or inside the progressive web app.

But building your progressive web app with AMP content containers means making a bigger commitment to AMP—and its constraints—than most organizations will feel comfortable with. Whatever limitations exist for AMP will be limitations placed on your website from that point forward, because your AMP pages and your website would be one and the same. I don't recommend it.

The fact that AMP exists at all speaks volumes about our failure to build fast websites by default. If web professionals had been focusing on performance from the start, AMP would have undoubtedly received less traction—there wouldn't have been a gap to fill.

But websites continue to be slow, and AMP *does* exist, and it will likely be part of your considerations. If you choose to build AMP pages, use them to make your progressive web app a better experience as well.

CREDENTIAL MANAGEMENT

Logging into websites remains a user experience nightmare. People often can't remember their username, their password, or both. Even if they do remember, the process of entering it—an input of special characters into a field that masks typing—can be a herculean task. Many people work around these challenges by using the same password on multiple websites, which means that their accounts are only as safe as the least secure website that stores their credentials.

Over the years, browser makers and technology companies have tried to address these hurdles. For instance, browsers have added secure password managers that can sync across devices, offer additional security suggestions, and automatically fill in credentials for known websites.

Federated login is another way that technology companies have attempted to solve the credential problem. Federated logins allow people to log into a website using credentials from other sites—typically larger organizations like Google, Facebook, or Microsoft. In addition to eliminating the need to remember separate credentials, federated logins are also more secure because the large technology companies have more resources devoted to securing their networks and detecting malicious activity. Not storing credentials also decreases the risk for site owners.

Both password managers and federated login are improvements to the login process. Websites should support federated login, and add the necessary autocomplete attributes to their forms so that browsers can autofill the correct information (http://bkaprt.com/pwa/07-06/). But even these solutions leave friction in the login process.

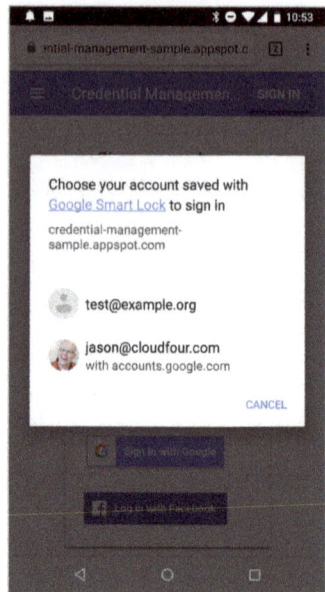

FIG 7.2: If someone has multiple accounts associated with a website, they will be asked which account they want to use. After they choose the account, they are signed in automatically without having to enter a password.

The Credential Management API eliminates this friction by automatically logging returning visitors into your website. It does this by building on top of the browser's password manager.

A website can ask if the user has stored credentials in the browser's password manager for the current site. If the user does have credentials stored, the website can ask to use those credentials. The browser will then display a small modal, allowing the user to confirm that they want to sign in using that account. If there are multiple accounts, the user can select which account to use (**FIG 7.2**). This works for both website-specific credentials and federated logins. After selecting an account, the user is logged in instantaneously.

Better yet, if someone returns to your website, you can log them in automatically using the same API (**FIG 7.3**). This only works if the user has previously acknowledged the auto-sign feature, used the Credential Management API in past, and didn't log out explicitly at the end of a previous visit.

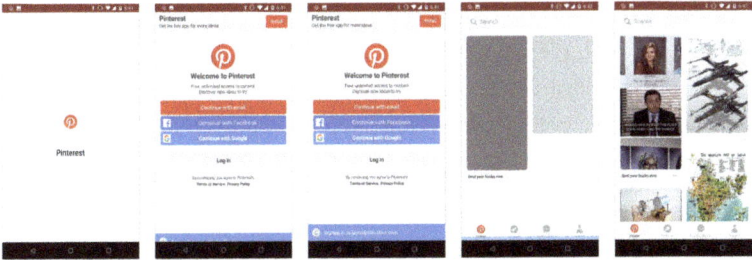

FIG 7.3: If you have previously logged in, Pinterest automatically logs you into the app without requiring any action from the user. The browser displays a brief notification to let users know they have been logged into the site.

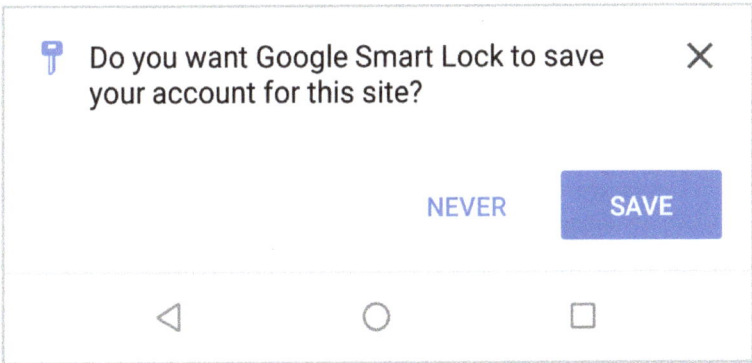

FIG 7.4: After a user creates an account, you can trigger a request to store the user's credentials in the browser's password vault.

You can also use the Credential Management API to ask people to store their credentials after they sign up for new accounts. Users will see a prompt to save their login in their password manager (FIG 7.4).

Automatic sign-in can significantly impact conversion. A study found that 92 percent of people say they have left a website when they couldn't remember their credentials instead of attempting to reset their password. A third of people say that they frequently leave websites because they cannot remember their credentials (http://bkaprt.com/pwa/07-07/, PDF).

Using the Credential Management API to simplify sign-in has made a difference for the companies that have implemented it. For instance, ecommerce company AliExpress saw an 85 percent drop in sign-in failures (http://bkaprt.com/pwa/07-08/). The Guardian saw a 44 percent increase in users who were signed in on two or more devices (http://bkaprt.com/pwa/07-09/).

The Credential Management API's main drawback is that Chrome, Opera, and UC Browser are the only browsers that currently support it. Apple has started development on the API, and both Mozilla and Microsoft have the standard under consideration for inclusion in their browsers.

But even with limited browser support, we can again look to progressive enhancement. All of the examples and instructions for using the Credential Management API assume a normal account sign-up and login flow that is enhanced only if the browser supports the API. In fact, there's no way to implement this API *without* using progressive enhancement, because you have to account for scenarios where a user declines to use their stored credentials. Progressive enhancement allows you to take advantage of the Credential Management API to provide a better login experience for those with browsers support, without leaving anyone else behind.

The Web Authentication API (also known as WebAuthn) is an extension to the Credential Management API to support biometric hardware or external authentication systems like Yubikey USB hardware (http://bkaprt.com/pwa/07-10/). Put a different way, WebAuthn will make it possible to log into sites using your fingerprint or facial recognition, depending on the hardware available on your device (http://bkaprt.com/pwa/07-11/). If this standard gains wide adoption, typing in passwords may become an uncommon occurrence.

A fast login process and staying logged in over multiple visits are conventions that people expect from their native apps—but they're just as viable on progressive web apps. We should spend the extra time to make the credentialing process as easy as possible by implementing autocomplete, federated login, and the Credential Management API.

PAYMENT REQUEST

Credentialing isn't the only friction-filled process online: for ecommerce websites, the checkout process typically involves tediously filling in billing, shipping, and payment information. A study from credentials company Jumio found that 56 percent of US adults have abandoned mobile shopping carts (http://bkaprt.com/pwa/07-12/), while the Baymard Institute found an average cart abandonment rate of 69 percent (http://bkaprt.com/pwa/07-13/). Streamlining the checkout process could help reduce this abandonment rate and increase sales.

That's where the Payment Request API comes in. It allows someone to check out faster using the credit cards and addresses they have stored in the browser (**FIG 7.5**). Depending on the browser, users may be able to complete their purchase by using their fingerprint alone.

The Payment Request API is currently supported by Edge, Chrome, Samsung Internet, and Opera browsers. It is in development for Firefox.

Apple developed a proprietary version for payment processing called Apple Pay on the Web (commonly shortened to Apple Pay) or Apple Pay JS (http://bkaprt.com/pwa/07-14/). However, the same day that Apple announced Apple Pay on the Web, Theresa O'Connor from the WebKit team sent an email to the W3C group working on the Payment Request API to let them know that Apple was interested in bringing what it had learned from Apple Pay back to the working group (http://bkaprt.com/pwa/07-15/). True to their word, support for the Payment Request API shipped in iOS 11.3. If you need to support older versions of iOS, Google has created a Payment Request API wrapper for Apple Pay (http://bkaprt.com/pwa/07-16/).

While the technical implementation of the Payment Request API is standardized, the requirements and user experience on each platform may differ. For example, Apple requires merchant validation to use Apple Pay so that access can be revoked if it is abused. Both Samsung and Apple allow for purchases to be approved via fingerprint, but not every browser supports fingerprint scanning yet. The good news is that, from a code

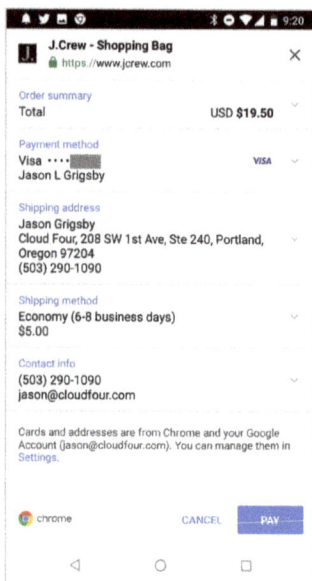

FIG 7.5: The Payment Request API provides a quick and easy way to use credit cards that users have stored on their device.

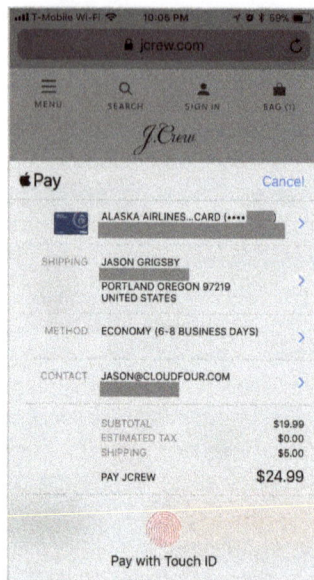

FIG 7.6: J.Crew's website uses the Payment Request API, which works on iOS as well as Android. Unfortunately, I forgot that taking screenshots on iOS requires pressing the home button—and since that's also the fingerprint sensor, I wound up with a pair of socks before I could capture this image.

perspective, you can use a single API and it will work the same, regardless of how the browser implements the API.

If you accept payments on the web, it's likely that implementing the Payment Request API will pay for itself. PureFormulas found that its progressive web app's "checkout process is 4x faster with half as many clicks" after implementing the Payment Request API (http://bkaprt.com/pwa/07-17/). Campmor saw a 65 percent decrease in cart abandonment and a 10 percent increase in completed checkouts (http://bkaprt.com/pwa/07-18/). J.Crew saw the average time spent in checkout decrease 75 percent after implementation (http://bkaprt.com/pwa/07-19/). I can personally testify to the speed and efficiency of J.Crew's checkout process—while trying to take screenshots for this book, I accidentally purchased their socks (FIG 7.6).

For browsers that don't yet support the Payment Request API, you should make sure that the forms on your checkout flow contain the autocomplete attributes necessary to help browsers autofill any stored addresses and credit cards the customer might have on file (http://bkaprt.com/pwa/07-06/). But with broad and enthusiastic support from browser makers for the Payment Request API, the days of users being forced to fill out lengthy forms to make purchases are numbered.

LEVEL UP

The web has grown up tremendously over the last few years and is more capable than it has ever been, but many people don't yet realize that this evolution has occurred. Therein lies the true superpower of progressive web apps—inspiring organizations to look at their website experiences in a whole new way.

As part of these conversations, it's inevitable that organizations will look at features that go beyond what is technically part of progressive web apps. Technologies like Payment Request, Credential Management, AMP, and other cool new browser features can support the sites we're reimagining with progressive web apps.

We should welcome these conversations, and not limit our progressive web apps to HTTPS, service workers, and manifest files. The biggest returns will come from combining technologies into new experiences that were never possible before.

8 PROGRESSIVE ROADMAP

THERE'S A LOT TO BE EXCITED ABOUT when it comes to progressive web apps, but it can also be a bit overwhelming. It can be difficult to see the path from a traditional website to a progressive web app that works offline and sends push notifications.

Don't despair! One of the hidden benefits of progressive web apps is that you can start right now and roll out features incrementally. That's what we did at Cloud Four when we built our progressive web app (FIG 8.1).

We knew we wanted to build a progressive web app, so when we redesigned our site, we launched it on HTTPS so we'd be ready. Then a couple of months later, we added a service worker with a simple fallback page. A bit after that, we added offline access to recently viewed pages and a notification when the network is unavailable. Push notifications were the last piece we added before we finally announced our progressive web app.

The best part about our incremental rollout is that at each step along the way, we made improvements to the experience for our visitors. We didn't have to wait until we had everything in place to show the benefits. We haven't stopped making improvements, either.

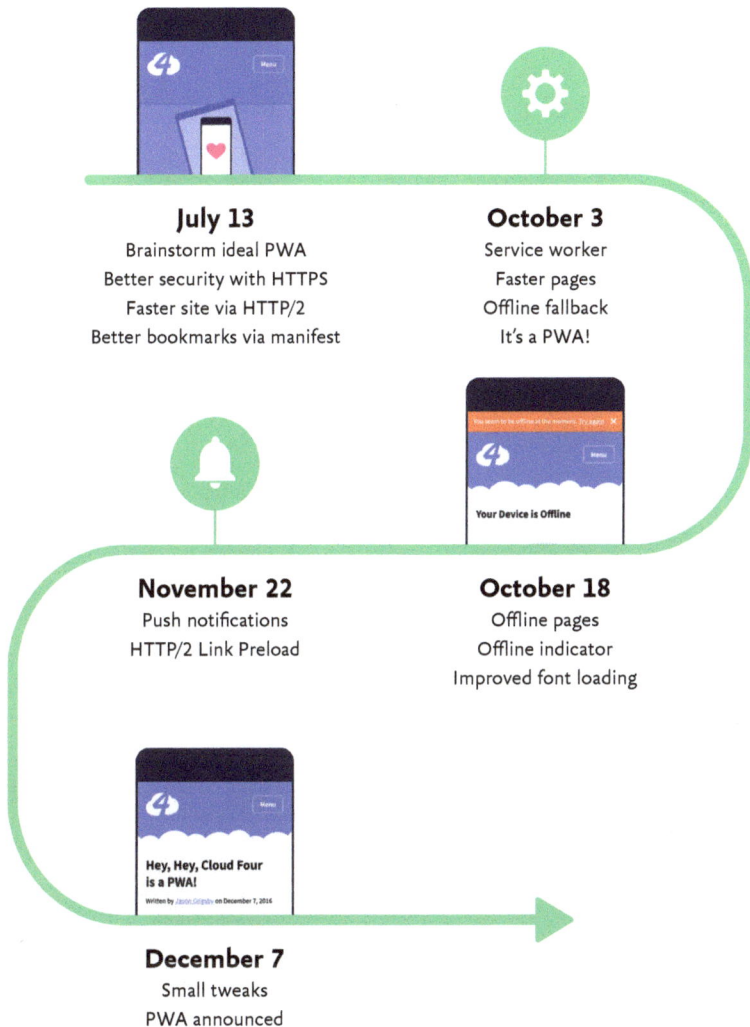

July 13
Brainstorm ideal PWA
Better security with HTTPS
Faster site via HTTP/2
Better bookmarks via manifest

October 3
Service worker
Faster pages
Offline fallback
It's a PWA!

November 22
Push notifications
HTTP/2 Link Preload

October 18
Offline pages
Offline indicator
Improved font loading

December 7
Small tweaks
PWA announced

FIG 8.1: At Cloud Four, we incrementally rolled out progressive web app features as we completed them.

You can build a roadmap for your progressive web app as well, focused on delivering incremental improvements. Each milestone should benefit your users in some way. Your plan can consist of four or five phases, depending on the current state of your website:

- Define your destination
- Address technical debt
- Build a baseline PWA
- Add front-end features
- Plan for the future

DEFINE YOUR DESTINATION

The first step in building your roadmap is figuring out where you want to go. Explore progressive web app features and determine which ones would make the most sense for your organization based on your user funnel and the core journeys you want to optimize.

Because we know there is no common definition of what a web app is, this is an opportunity to gauge what others in your organization have in their minds when they think of an "app." Building alignment around definitions and goals is key.

At this point in the process, don't constrain yourself with what you think is currently possible. If you think push notifications integrated into your core product would be a huge win for your customers, write it down. It may take you months to get to that point, but at least you'll know what you're building towards.

This is a great opportunity to bring more stakeholders together to talk about progressive web apps and what makes sense for your organization. Make this a fun exercise where you can talk about possibilities without being fettered by day-to-day realities. Even pie-in-the-sky ideas can be put on the roadmap (albeit far out in the future) so they don't derail you later in the process. Giving everyone a chance to feel heard will also help later when you need to get buy-in on specific initiatives.

Keep in mind that there may be opportunities to use other web standards that aren't technically part of progressive web apps. If you manage an ecommerce website, there's a good chance that utilizing the Payment Request API will be on your list of ideal features for your progressive web app.

When brainstorming, consider some of the factors we've covered in the previous chapters:

- How much do you want it to feel like a native app?
- How important is installation and discovery?
- How would offline access benefit your users?
- How can you use push notifications to reengage users?
- Would your users benefit from additional features beyond PWAs?

For each, there is a continuum from the simplest version to the most complex version. How would each of these impact your most important metrics? Which ones would have the largest impact? Where would your ideal progressive web app land on the continuum for each factor (FIG 8.2)?

After everyone has brainstormed what the progressive web app can be, then you can start to assess priorities. We're fans of using the KJ-Method, an efficient and democratic process for identifying, grouping, and voting on the most important features (http://bkaprt.com/pwa/08-01/).

Don't get stuck on the prioritization. The goal here isn't to determine the order on your roadmap—that will be more influenced by the realities of technical implementation and the needs of the user experience. What you want by the end of this planning phase is consensus around what your ideal progressive web app would look like so you have a destination to march towards.

ADDRESS TECHNICAL DEBT

Every website could benefit from becoming a progressive web app, but there's a chance that some housekeeping may be in order before you're ready to start down the road.

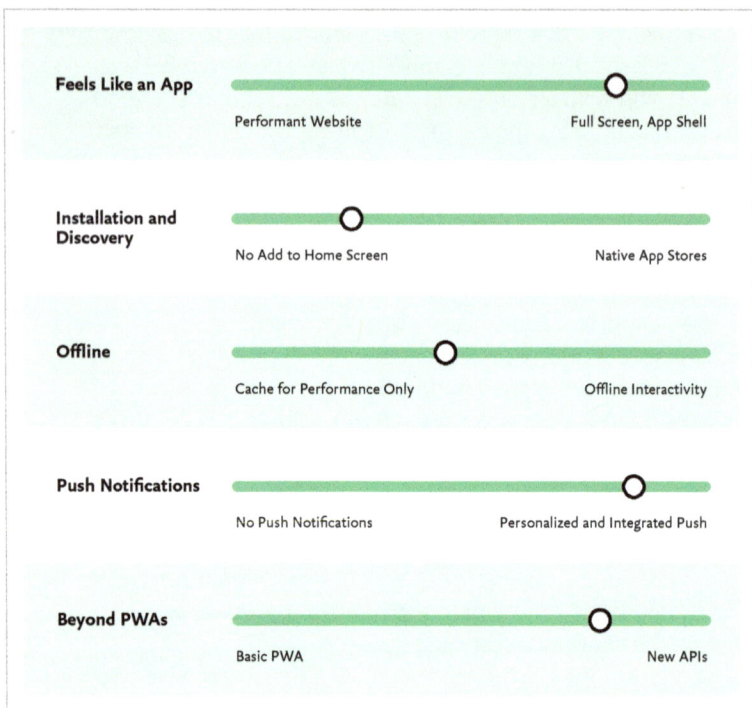

FIG 8.2: The dots in the chart illustrate what the ideal progressive web app for an ecommerce company might look like.

In particular, your current website needs to meet a reasonable performance and usability level. If your current website takes thirty seconds to load on a fast 3G connection, then adding a service worker isn't going to be sufficient to get the full return on investment that other companies are seeing with their progressive web apps. Remember, service workers only have an impact on the second page load. How many people will get to the second page if the first page takes too long to render?

There are a number of tools you can use to test your website performance. Lighthouse, available in the Audits tab inside Chrome DevTools, will test your progressive web app for performance, accessibility, and other web best practices (FIG 8.3).

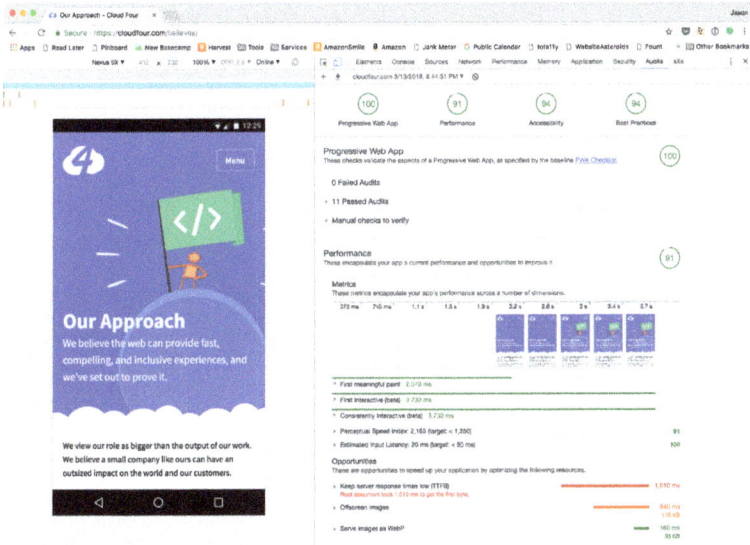

FIG 8.3: Lighthouse, found in the Audits tab in Chrome DevTools, will audit your website and suggest areas for improvement.

It's also available as a command line tool that you can integrate into your development workflow.

WebPagetest is a free service that lets you run performance tests from devices all over the world (http://bkaprt.com/pwa/08-02/). You can test on real mobile devices or simulate devices. I recommend using mid-tier mobile devices on a simulated fast 3G network as your baseline test (**FIG 8.4**). That's the same benchmark used in Lighthouse, which means that it will be easier to compare results.

Another tool, webhint, works a little differently than Lighthouse and WebPagetest. Webhint is an open-source linting tool that can check your website from a variety of different perspectives including performance, accessibility, security, and progressive web app support (http://bkaprt.com/pwa/08-03/). Webhint doesn't give you a grade like WebPagetest and Lighthouse, but raises issues and points you to the relevant documentation describing why something might be a problem. You

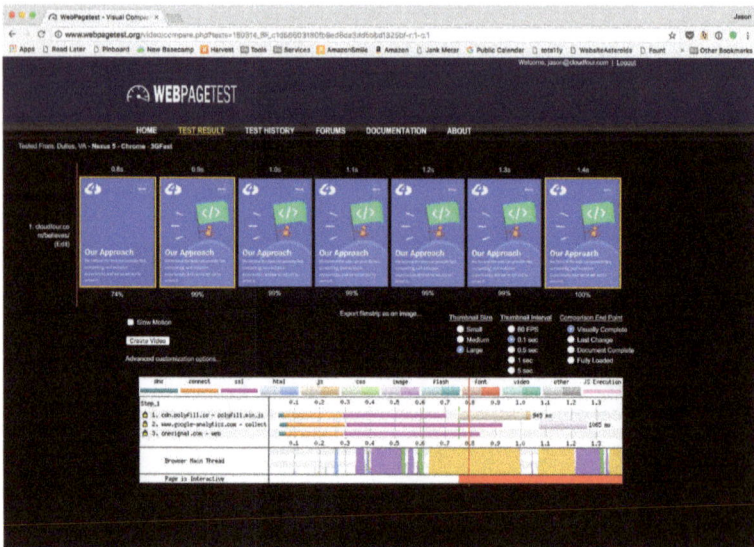

FIG 8.4: WebPagetest is a remarkable, free service for performance testing. Bookmark it and use it frequently.

can use this information to gauge for yourself how, or even whether, to resolve the issues.

In addition to performance assessments, you should also look at how well your website responds to different screen sizes and inputs. If your website isn't a responsive design or isn't touch-friendly, you may need to address these issues before you start on your progressive web app.

More than once at Cloud Four, we've had to help organizations clear some of their technical debt before we could build their progressive web app. Sure, you could add a service worker and a manifest file and meet the technical requirements of a progressive web app, but if the site is exceedingly slow and difficult to use, then simply adding a service worker isn't going to make a big difference.

Remember, the goal isn't to create a progressive web app. The goal is to make a better experience for your users and meet your business objectives.

Benchmarking

While you're evaluating your current site performance, it's a good time to start thinking about how you're going to measure—and maintain—the improvements you're going to see from the progressive web app. You'll want to record benchmarks that you can use to evaluate the improvements.

Consider signing up for performance monitoring tools like SpeedCurve, Calibre, or Akamai as a way to automate performance testing. Look for key performance indicators that can be tracked in analytics to tie your improvements to your company's bottom line. If you don't have a key metric in mind, consider First Input Delay (FID) and Time to Interactive (TTI) (http://bkaprt.com/pwa/08-04/). These metrics are designed to assess how quickly your site becomes responsive to input.

Once you've determined your key metrics, set a performance budget for your website—and stick to it (http://bkaprt.com/pwa/03-22/).

Finally, please keep good notes and write about the results of your work. Our industry improves when people share what they learn (and we'd love to highlight your company's success on PWA Stats!).

BUILD A BASELINE PWA

Once your website is in a decent starting place, it's time to build a baseline progressive web app that is secure, uses a service worker for performance improvements, and has an offline fallback page. It may not be everything you've dreamed of yet, but it is a good way for your team to get acquainted with the technologies that make up progressive web apps.

A baseline progressive web app can typically be completed without disruption to end users and without the assistance of other groups in your organization.

Creating a manifest file

The first milestone on your path should be to add a manifest file to your website. Manifest files are short and easy to create. In many ways, they're hardly worth their own milestone.

However, manifests are also foundational—they're what many browsers look at to determine what icons to use when someone bookmarks your website, and whether you have a progressive web app or not. Absent a manifest file, if someone bookmarks your site on an Android device, the icon they see will be a plain gray square with the first letter of your site's domain in it. Not exactly the best presentation of your brand.

This is a no-brainer. Create a short manifest file and some icons to go with it. PWA Builder can generate a manifest file for you automatically (http://bkaprt.com/pwa/04-09/). Look back to Chapter 4 if you want to tweak some of the settings. If someone is loyal enough to bookmark your site before your PWA is available, reward them with a branded experience.

Turning on HTTPS

Turning on HTTPS for your website should be the next stop on your journey. Depending on your site infrastructure, this milestone can be a mere formality or a major initiative.

For the Cloud Four website, turning on HTTPS meant flipping a switch in our hosting provider's admin console. The provider provisioned the certificate from Let's Encrypt and installed it automatically.

But not all sites are so straightforward—one of our clients took nearly six months to move to HTTPS because of the complexity of its site. When you move to HTTPS, you have to make sure *everything* on every page is under HTTPS, or your users will see mixed content warnings. Tracking down assets and third-party code on a large website can take some time.

While you're at it, look to see if you can move from HTTP to HTTP/2 at the same time. Our hosting provider automatically gives HTTP/2 to sites on HTTPS. Some of our clients use Akamai's content delivery network and have also seen an automatic switch to HTTP/2 once on HTTPS.

HTTP/2 handles connections to servers differently. You may find that some of the things you did previously to optimize your sites for HTTP/1 are no longer necessary on HTTP/2. For example, many sites distribute images across multiple subdomains—such as 1.example.org, 2.example.org, 3.example.org—as a way to bypass HTTP/1's limit on the number of concurrent connections. This technique, called *domain sharding*, doesn't make sense for HTTP/2, and may actually slow things down (http://bkaprt.com/pwa/08-05/).

As soon as you can deploy HTTPS, you should—there's no need to wait for other features on your roadmap. HTTPS will benefit your users immediately by providing your site over a secure connection. And if you're able to launch HTTP/2 as well, they'll also get a faster site.

Adding a service worker

Perhaps your ideal progressive web app includes precaching pages, background sync, and other advanced features. The full offline experience you envision will take more time to build—but in the short run, you may be able to use service workers for performance and reliability improvements.

Most of the more complex offline scenarios build on top of a performance baseline in some way. There will likely be files you can cache to speed up the site, or areas of the site that don't require offline access but can use the performance boost.

Feel free to leverage existing tools to speed up the construction of your service worker. Workbox is an open-source library maintained by Google that supports many common caching scenarios (http://bkaprt.com/pwa/05-01/). And, again, you can use PWA Builder to create a basic service worker. Tools like these will help you get a simple service worker up and running quickly.

Even with the most basic of service workers, your customers will likely see a noticeable speed improvement. Be sure to measure how much improvement you see from these caching changes; compare the speed of repeat page views—where the service worker has been installed—against your benchmark.

Once you've installed your service worker, you can also add an offline fallback page. You don't need the fallback page to ship the service worker, so don't hold up—our goal with this roadmap is to deliver value at every opportunity possible.

Preparing for discovery

A quick word of caution: depending on the browser, users may start to see ambient badges or prompts as you deploy portions of your baseline experience. Firefox will start displaying the badge as soon as your app is on HTTPS and has a manifest file. Other browsers will wait until you have a service worker installed. For more specifics on the criteria browsers use, see Chapter 4, but keep in mind that the criteria are constantly evolving.

If you're okay with users seeing badges or prompts to install your app, then there is nothing for you to do. However, if you're not ready for people to install your progressive web app, or are trying to keep a low profile for your project, there are ways to prevent the browser from prompting people.

The easiest solution is to change the display mode setting in your manifest to `browser`. Firefox is the only browser that will display an add-to-homescreen badge if the display mode is set to `browser`, so that's the quickest way to reduce the number of people who are prompted.

You could also do what Flipkart did and suppress the add-to-homescreen prompts by intercepting the `BeforeInstallPrompt` event and delaying it until you're ready. You can log the event to your analytics so you understand where and when people typically hit the user engagement threshold, which will help you decide later where and when you want to trigger the prompts.

In most cases, there's no harm, and likely some benefit, in having people install your progressive web app even if it isn't yet everything you hope it will be. That's the appeal of building a progressive web app incrementally.

ADD FRONT-END FEATURES

After getting the basics out of the way, picking which features to work on next can be daunting. Features like push notification integration or the Payment Request API might have the largest impact on your organization, but they will also likely involve multiple teams inside your company.

The more teams that are involved in the development of a feature, the more time it will require. Instead, I recommend first focusing on features that can be accomplished by the same small team that worked on your baseline progressive web app. Racking up those front-end wins will help you maintain momentum as you move forward.

Improving service workers

The first area to explore is making additions to your service worker to support more offline capabilities—most improvements can be done without worrying about the impact on other parts of your organization and infrastructure.

The lowest of low-hanging fruit here is caching recently viewed pages. You can do this without making any changes to the user interface or backend, though you may want to consider adding a notification alert if someone is viewing a page that is offline.

Other offline functionality, like precaching pages and handling offline interactivity, can also be handled primarily on the front end. They're perfect candidates for your next milestones.

Adding push notifications

Another quick win for the front-end team could be the addition of simple push notification support. Some third-party push notification services offer integration with content management systems. For example, several services provide WordPress plugins that allow you to send notifications when you publish new articles.

There's nothing wrong with using a push notification service for simple use cases like this, so long as you're not spamming people with requests to sign up for notifications. As we discussed in Chapter 6, it's important to be respectful of your audience and to ask them for permission to send push notifications when they're most likely to accept. Don't abuse their privacy and trust.

Basic notification services aren't going to be as powerful as any messaging program you put together that integrates with your internal systems. Personalized messages are far more effective, but also require a larger integration effort.

If your long-term roadmap includes push notifications, start evaluating the third-party providers early on. If you identify a provider that you want to work with, and the service can quickly provide value to users, then go ahead and implement. Setting up simple push notifications can help your team get a better understanding of the integration needed for your larger initiatives.

To be clear: don't implement push notifications just for the sake of having push notifications. Not only will you irritate your users, but you'll likely damage your credibility inside your organization. Given the choice between a quick win on push notifications or taking more time to determine how push notifications can enhance an important user journey, always choose the latter.

PLAN FOR THE FUTURE

As you build your roadmap, you'll likely identify features that will take more time, have more uncertainty, or require the cooperation of more teams in your organization. By necessity, these items will end up further out on your roadmap, but that doesn't mean that you should ignore them during the early phases of your project.

While you're developing the early portions of your progressive web app, set aside time to research these features. Meet with the teams that you'll need assistance from to find out what the implementation would truly require and any hidden challenges you may encounter.

If you have a native app, find out what features it provides that causes your organization to steer people to the native app. Those features are likely ones that have more value to your organization and are places where getting the feature in front of a larger audience on the web could make a significant difference.

Learn more about other initiatives within your company that could dovetail with your efforts. For example, moving from a traditional website architecture to an app shell model is a major undertaking. It may not make sense to undertake that transition for your progressive web app alone. However, many organizations have been moving toward API-centric architectures and JavaScript on the server—particularly node-based infrastructures—for reasons unrelated to progressive web apps. You may find that the team responsible is already making plans to move to an architecture that would be more conducive to your PWA aspirations.

Be prepared: some of the thornier issues may not be technological. Moving to the Payment Request API, for example, requires changing the way payment is accepted by your organization, and the agreements about what types of payment are accepted are more likely to be governed by business considerations than technical ones.

As you start putting together your plans for these larger initiatives, you may find that you have less control over the timing of them than you would like. There's not much you can do about it if one of your features is dependent on an infrastructure change that won't be complete for several months.

But to the degree that you have control over the relative timing and priority of these initiatives, prioritize the items that you believe will have the biggest impact on the company's bottom line. Doing so will help you maintain momentum.

DELIVER INCREMENTAL IMPROVEMENTS

When we put all those steps together, we get a comprehensive timeline of all the large and small initiatives and PWA priorities, vetted by all stakeholders and across all teams (FIG 8.5). And while we started our journey by deciding what destination we wanted to reach, we've delivered value for our users and our organization at multiple points along the way.

By delivering value incrementally and documenting the impact of each improvement—you *are* measuring the improvements, aren't you?—you'll build up credibility in your organization. There's nothing that removes roadblocks more quickly than demonstrating how the work you're doing improves the experiences for your customers and impacts the bottom line.

The remaining features on your roadmap will likely require the cooperation of other parts of your organization. When you start work on the larger initiatives, you can cash in on some of the credibility you've earned along the way to make sure the work you need from other teams is high on their list as well.

One of the great things about a progressive web app is that, unlike a native app, you can continually improve on it and ship new changes. You don't have to wrap up all of your code, build a binary, submit it to an app store, and wait for approval. You can make as many changes as you want, as frequently as you want.

Use that flexibility to your advantage by putting together a plan that rolls out improvements on an ongoing basis. There's nothing wrong with starting small and building up to your dream progressive web app experience. In fact, securing small wins on the way toward your ideal progressive web app is the smart way to ensure your organization—and your users—experience the most value.

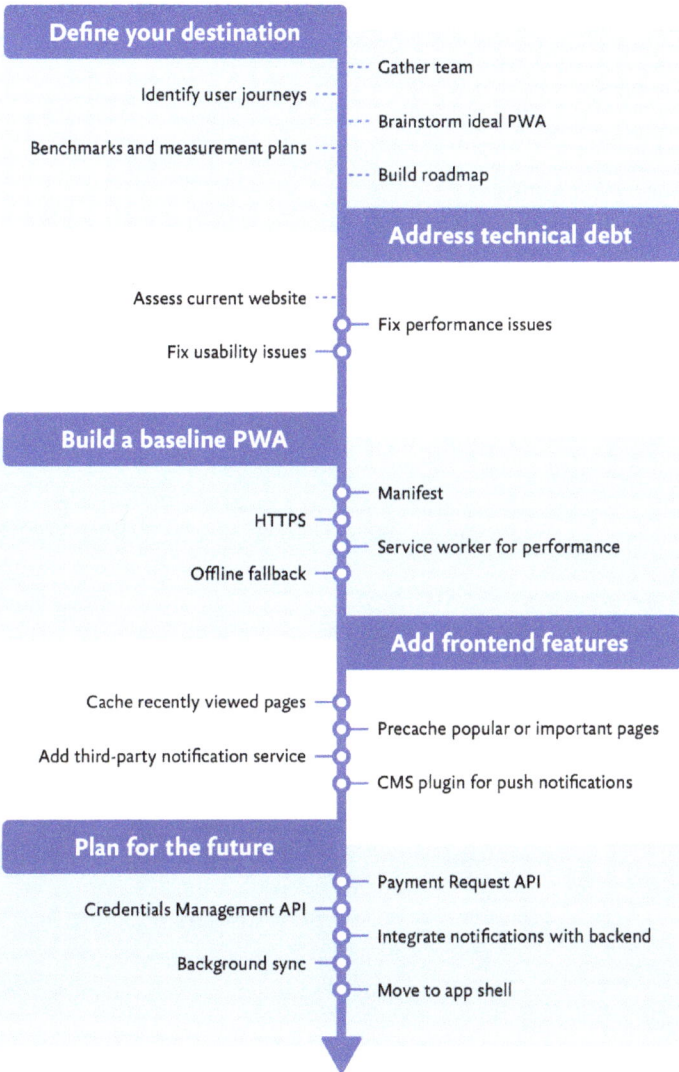

Define your destination
- Gather team
- Identify user journeys
- Brainstorm ideal PWA
- Benchmarks and measurement plans
- Build roadmap

Address technical debt
- Assess current website
- Fix performance issues
- Fix usability issues

Build a baseline PWA
- Manifest
- HTTPS
- Service worker for performance
- Offline fallback

Add frontend features
- Cache recently viewed pages
- Precache popular or important pages
- Add third-party notification service
- CMS plugin for push notifications

Plan for the future
- Payment Request API
- Credentials Management API
- Integrate notifications with backend
- Background sync
- Move to app shell

FIG 8.5: A high-level roadmap for a fictional ecommerce company that wants to build a progressive web app with app shell, Payment Request API, Credential Management, and sophisticated push notifications. Each dot on the timeline represents a potential release point where an improvement could be delivered to end users.

9 A WEB FOR EVERYONE

EARLIER, I DISMISSED THE NOTION that progressive web apps are in competition with native apps. I firmly believe that, for most organizations, that holds true: the decision of whether to build a progressive web app or not has little to do with whether or not the organization already has a native app.

But even if progressive web apps and native apps aren't in direct conflict, progressive web apps were created in part as a response to perceived threats to the web: mobile app stores and their gatekeepers, less attention (and budget) given to the mobile web, bloated and inaccessible websites... As we've been slow to fix these problems, we've lost ground to native apps.

Since its inception, people have declared the web endangered or even dead (**FIG 9.1**) (http://bkaprt.com/pwa/09-01/). In 2016, Alex Russell tweeted that "the reality [is] that the web is in crisis. Actual, real, serious crisis" (http://bkaprt.com/pwa/09-02/). Bruce Lawson, who was working for Opera browser at the time, agreed that Russell wasn't "being alarmist; it's true" (http://bkaprt.com/pwa/09-03/).

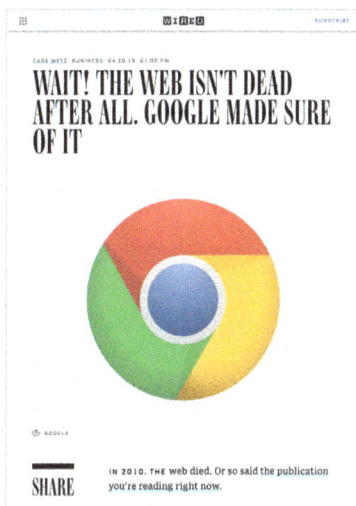

FIG 9.1: In 2010, Wired magazine declared the web dead (left). In 2016, Wired said the web had been saved by progressive web apps (right). The truth lies somewhere in between.

I worry about these threats, too—but I also recognize that, despite these threats, the mobile web is growing twice as fast, and is three times the size of native app audiences (http://bkaprt. com/pwa/09-04/). The day after Alex issued a clarion call to save the web, Recode's big headline was that "The App Boom is Over" (http://bkaprt.com/pwa/09-05/). Developer Scott Jehl summed up the confusion well (http://bkaprt.com/pwa/09-06/):

Looks left: "the web is dead"
Looks right: "apps are dead"
...
Looks up the weather forecast

There's no denying that progressive web apps came about as a way to put the web back on equal footing with native development. As we look to the future of the web, perhaps we need to consider building progressive web apps as part of our responsibility to it.

PASCAL'S WAGER

On most days, I feel bullish about the future of the web for all of the reasons I've talked about in this book. Occasionally, news breaks—like the announcement of Android Instant Apps, which promised to replicate the instant access of the web—and I worry about its future again (http://bkaprt.com/pwa/09-07/).

But we don't have to determine whether the web is facing a crisis or not to know the way forward. If we borrow Pascal's Wager—a decision-making matrix created by French philosopher Blaise Pascal—and apply it to the hypothesis that the web is threatened and progressive web apps could save it, there are four possible outcomes:

- If the web isn't threatened and we don't build PWAs, nothing changes.
- If the web isn't threatened and we build PWAs, users have a faster and better web experience.
- If the web *is* threatened and we build PWAs, users have a faster, better web experience, *and* we save the web.
- If the web *is* threatened and we don't build PWAs, the web loses to native.

Whether the web is truly threatened or not, the best course of action is to build progressive web apps. The worst that can happen if we build progressive web apps is that people have a better web experience. The worst that can happen if we don't is…well…the worst.

The rationale for building a progressive web app described in Chapter 2 remains the same. Few companies are going to invest in building progressive web apps for altruistic reasons; they're going to build them because they can lead to increases in conversion, engagement, and revenue.

And if we happen to help save the open web along the way, that's cool too.

THE NEXT BILLION WEB USERS

There are over 1.3 billion people living in India, but only 27 percent of India's population has access to the Internet. Similarly low levels of internet access exist in other Asian and African countries. Because of this, it is commonly said that the next billion internet users will come from these emerging markets.

The people who are coming online in these emerging markets share some characteristics. They are likely to be mobile-only internet users. They often have slow (2G or 3G) and unreliable network connections. They are typically on metered connections and have to pay per megabyte for data transfers. They have limited storage on their phones and are forced to limit the number of apps they install.

Given these factors, it shouldn't be a surprise that the earliest adoption and most innovative use of progressive web apps is happening in countries like Kenya, Nigeria, Indonesia, and India. Building a fast, light web experience is helping companies there reach new customers and recover customers they had lost. Amar Nagaram, vice president of engineering for India-based retailer Flipkart, said that "around 60% of visitors to Flipkart's mobile web had uninstalled the [native] app earlier, primarily due to lack of device storage space" (http://bkaprt.com/pwa/09-08/). And Indian cab aggregator Ola found that 20 percent of the people who booked a ride using their progressive web app had previously uninstalled their native app (http://bkaprt.com/pwa/09-09/).

Avoiding apps because of storage concerns isn't a behavior limited to emerging markets. One of our clients, a large U.S.-based retailer, recently told us about their lack of success in convincing customers to install their native app. When they investigated further, customers told them they were worried about how much space it would take up. Their worry was justified—the native app weighed in at around 100MB.

Progressive web apps are not only a potential way to save the web—they also represent the best way to reach the next billion internet users.

ESCAPING TABS WITHOUT LOSING OUR SOULS

Since the iPhone app store launched in 2008, web developers and browser makers have been chasing native apps in a foolhardy way. Their attempt to combat a collective inferiority complex has led to well-intentioned solutions that have abandoned the web's inherent strengths, trading ubiquity and linkability for the promise that something would "feel native."

The article in which Frances Berriman and Alex Russell coined progressive web apps had a subtitle: "Escaping Tabs Without Losing Our Soul." The key difference between progressive web apps and what has come before them is that progressive web apps aren't trying to replicate native; they aren't casting aside what makes the web great, but embracing it.

With progressive web apps, we can make our experiences available anywhere. They work from inside the browser, from a mobile homescreen, or in response to a push notification. They work on slow connections, and even offline. Any device with a browser can use a progressive web app.

More than anything, progressive web apps challenge us to reimagine what is possible on the web. We can build experiences that feel as rich, fast, and powerful as native apps, but with even greater reach. And because progressive web apps are built using the web, they are available regardless of operating system, device, or location. With progressive web apps, the web truly is for everyone.

ACKNOWLEDGMENTS

No amount of praise or thanks will sufficiently describe Lisa Maria Martin's impact on this book. From helping me hone my ideas into a coherent narrative to reworking my gibberish into intelligent sentences, Lisa Maria transformed everything she touched. For days after receiving her first line edit of the book, I couldn't stop telling people how much better the book was and how I had nothing to do with it. Lisa Maria is a godsend.

Katel LeDû has shepherded me from the earliest book pitch through to the very end of the book publishing process. My jaw dropped when I realized that, in addition to being A Book Apart's CEO and star podcaster, she also served as the copyeditor for this book. She is a multitalented impresario.

Thanks to Jeffrey Zeldman, Jason Santa Maria, and the rest of the A Book Apart crew for having faith in me and for making the book look wonderful. It is such a relief to know that the book's design was in capable hands.

Special thanks to Alex Russell and Aaron Gustafson for their detailed tech reviews. To Frances Berriman and Alex for their extremely generous foreword. To Jeremy Keith, Sarah Drasner, and Karolyn Hart for their early reviews and kind words.

Thanks to Owen Campbell-Moore, whose writings and conversations about progressive web app UX influenced my research in that area. To Rachel Nabors, Val Head, and Sarah Drasner for showing us how to use animation on the web. And to my extended family at An Event Apart—particularly Jeffrey Zeldman, Eric Meyer, Toby Malina, and Marci Eversole—for giving me the space to explore these ideas in front of an audience.

Thanks to all of the exceptional people working hard every day to make browsers and the web better: Dion Almaer, Rachel Andrew, Jake Archibald, Rey Bango, Andreas Bovens, Peter Beverloo, Jeff Burtoft, Chris Coyier, Ada Rose Cannon, Ben Galbraith, Matt Gaunt, Jenny Gove, Dean Hume, Paul Kinlan, Eiji Kitamura, Mustafa Kurtuldu, Chris Love, Pete LePage, Patrick Meenan, Peter O'Shaughnessy, Addy Osmani, Jeff Posnick, Bryan Rieger, Stephanie Rieger, Jennifer Robbins, Jenn Simmons, Steve Souders, Estelle Weyl, and many, many more people too numerous to name. You inspire me with your willingness to share what you've learned.

I want to particularly thank the developers of Flipkart, Wego, Trivago, and other early progressive web apps being made in Africa, India, and other parts of the world where the web is less reliable. You are guiding the way for the rest of us. Thanks to Bruce Lawson, Tal Oppenheimer, Tim Kadlec, and others who help remind us that the web should be accessible everywhere and to everyone.

Thanks to my coworkers and clients at Cloud Four whose projects and work inspired this book. Gerardo Rodriguez, and Lyza Gardner and Erik Jung before him, was crucial in exploring service workers and helping me understand what was possible. Tyler Sticka's illustrations in the book make complex topics seem simple. Megan Notarte's encouragement and assistance kept me going when work and life overwhelmed. And none of this would be possible without Aileen Jeffries, who started Cloud Four with me and has been a coworker or business partner for eighteen years now. Thank you.

Finally, I owe the largest debt of gratitude to my family and friends who were so understanding and supportive over the last year when I skipped events to work on the book. Even now, I write this while on vacation in Sunriver, Oregon. My wife, Dana, has taken our children, Katie and Danny, to the swimming pool while I stayed behind to finish the last bits of the book. There is no way this book would have happened without Dana and the rest of our family. I love you all and am looking forward to spending more time with you in the coming months.

RESOURCES

Were this a few decades earlier, this would be the point where I shuffle into the room barely able to carry a precariously balanced stack of books, loudly drop them on the table in front of you, and proclaim, "Read these next." Thankfully, the hyperlinks I'm going to share won't strain either of our backs.

Code

This book was intentionally devoid of code, but at some point, you'll need to learn how to write your own service workers and other PWA goodies.

- Jeremy Keith's *Going Offline* is a fantastic introduction to service workers and progressive web apps. written so that even non-developers can follow along (http://bkaprt.com/pwa/10-01/).
- *Progressive Web Apps* by Dean Alan Hume is another good introduction written for a developer audience (http://bkaprt.com/pwa/10-02/).
- Cloud Four cofounder Lyza Gardner detailed how she applied a service worker to her site in the aptly titled "Making a Service Worker" (http://bkaprt.com/pwa/10-03/).
- Matt Gaunt's *Web Push Book* covers all you need to know about both the code and the user experience of push notifications (http://bkaprt.com/pwa/10-04/).
- The Microsoft Edge team has built a zodiac-themed website that demonstrates how push notifications work, including all of the code necessary to send astrology trivia on a daily basis (http://bkaprt.com/pwa/10-05/).

User Experience

- Mustafa Kurtuldu's "Offline UX Considerations" provides useful guidance and examples (http://bkaprt.com/pwa/05-03/).

- "Designing Great Progressive Web Apps" by Owen Campbell-Moore challenges us to adopt some of the best UX features of native apps (http://bkaprt.com/pwa/03-02/).
- Campbell-Moore also wrote "Best Practices for Push Notifications Permissions UX," which we used to guide us when we implemented push on Cloud Four's website (http://bkaprt.com/pwa/10-06/).
- Sarah Drasner's inspiring "Native-Like Animations for Page Transitions on the Web" made me want to redesign some of the progressive web apps we've built (http://bkaprt.com/pwa/03-19/).

Case Studies

- Google publishes periodic case studies on the Google Developer site (http://bkaprt.com/pwa/10-07/).
- Addy Osmani has written detailed performance case studies on Pinterest, Tinder, Treebo, React, and others on his Medium account (http://bkaprt.com/pwa/10-08/).
- We keep track of case studies at PWA Stats. If we've missed something, please let us know (http://bkaprt.com/pwa/02-05/).

Beyond PWAs

- If you have to use AMP, "Bootstrapping Progressive Web Apps with amp-install-serviceworker" by Ruadhán O'Donoghue tells you how to use your AMP pages to make your PWA faster (http://bkaprt.com/pwa/10-09/).
- Eiji Kitamura and Meggin Kearney describe how to implement the Credential Management API (http://bkaprt.com/pwa/10-10/).
- Peter O'Shaughnessy's how-to for Smashing Magazine is an excellent introduction to the Payment Request API (http://bkaprt.com/pwa/10-11/).
- "Introducing the Web Share API" by Paul Kinlan and Sam Thorogood is your one-stop shop for everything you need to know about this API (http://bkaprt.com/pwa/03-14/).

Tools

- I found Peter Beverloo's Notification Generator both useful and fun for testing different types of push notifications (http://bkaprt.com/pwa/06-05/).
- Google maintains a JavaScript library called Workbox that makes service worker development easier (http://bkaprt.com/pwa/05-01/).
- WebPagetest is my go-to resource for testing page performance (http://bkaprt.com/pwa/08-02/).
- Lighthouse, a progressive web app testing tool, is available in Chrome DevTools, or can be run from the command line or as a Node module (http://bkaprt.com/pwa/10-12/).
- PWA Builder by Microsoft creates icons, manifest files, and service workers, and helps you submit your app to the Microsoft Store (http://bkaprt.com/pwa/04-09/).
- For a different perspective on the web, test your site using Tim Kadlec's What Does My Site Cost tool to see what people in other countries will pay as a portion of their income to download your site (http://bkaprt.com/pwa/10-13/).

Impact

Finally, I want to leave you with a couple of articles that broadened my view and reminded me why progressive web apps matter.

- Alex Russell questions conventional wisdom about native apps and their impact on businesses in "Why Are App Install Banners Still A Thing?" (http://bkaprt.com/pwa/02-09/).
- Bruce Lawson's two-part series entitled "World Wide Web, Not Wealthy Western Web" provides a global perspective on where the web is headed and the role progressive web apps play in that future (http://bkaprt.com/pwa/10-14/).
- In a similar vein, Tal Oppenheimer's presentation "The Web for the Entire World" is an eye-opening look at how people in other countries experience the web, and how we can build a web that works better for everyone (http://bkaprt.com/pwa/10-15/).

REFERENCES

Shortened URLs are numbered sequentially; the related long URLs are listed below for reference.

Introduction

00-01 https://developers.google.com/web/showcase/2016/housing

00-02 https://www.youtube.com/watch?v=_pmjBZi5zY0

00-03 http://www.fastmoving.co.za/news/retailer-news-16/lancome-speeds-its-mobile-site-with-google-s-progressive-web-apps-9959

Chapter 1

01-01 https://infrequently.org/2015/06/progressive-apps-escaping-tabs-without-losing-our-soul/

01-02 https://developers.google.com/web/progressive-web-apps/

01-03 https://www.youtube.com/watch?v=PsgW-0M67TQ&t=340

01-04 https://fberriman.com/2017/06/26/naming-progressive-web-apps/

01-05 https://infrequently.org/2015/06/progressive-apps-escaping-tabs-without-losing-our-soul/

01-06 https://adactio.com/journal/13098

01-07 https://infrequently.org/2015/06/progressive-apps-escaping-tabs-without-losing-our-soul/

Chapter 2

02-01 https://magento.com/news-room/press-releases/magento-reimagine-mobile-commerce-progressive-web-app

02-02 https://www.youtube.com/watch?v=Di7RvMlk9io

02-03 https://medium.com/dev-channel/a-pinterest-progressive-web-app-performance-case-study-3bd6ed2e6154

02-04 https://youtu.be/PsgW-0M67TQ?t=34m4s

02-05 https://www.pwastats.com

02-06 https://www.comscore.com/Insights/Presentations-and-Whitepapers/2016/The-2016-US-Mobile-App-Report

02-07 https://www.comscore.com/Insights/Presentations-and-Whitepapers/2017/The-2017-US-Mobile-App-Report

02-08 https://www.emarketer.com/Article/Cost-of-Acquiring-Mobile-App-User/1016688

02-09 http://andrewchen.co/new-data-shows-why-losing-80-of-your-mobile-users-is-normal-and-that-the-best-apps-do-much-better/

02-10 https://medium.com/dev-channel/why-are-app-install-banners-still-a-thing-18f3952d349a

02-11 https://letsencrypt.org

02-12 https://istlsfastyet.com/

02-13 https://blog.chromium.org/2017/04/next-steps-toward-more-connection.html

02-14 https://www.chromium.org/Home/chromium-security/marking-http-as-non-secure

02-15 https://blog.mozilla.org/security/2018/01/15/secure-contexts-everywhere/

02-16 https://www.doubleclickbygoogle.com/articles/mobile-speed-matters/

02-17 https://www.slideshare.net/devonauerswald/walmart-pagespeedslide

02-18 https://medium.com/@addyosmani/a-tinder-progressive-web-app-performance-case-study-78919d98ece0

02-19 https://tech.treebo.com/we-didnt-see-a-speed-limit-so-we-made-it-faster-treebo-and-pwas-the-journey-so-far-f7378410abc7

02-20 https://developers.google.com/web/showcase/2016/extra

02-21 https://developers.google.com/web/showcase/2017/lancome

02-22 https://developers.google.com/web/showcase/2017/olx

02-23 https://developers.google.com/web/showcase/2016/carnival

02-24 https://www.mobilemarketer.com/news/can-progressive-web-apps-solve-the-app-vs-browser-dilemma/510344/

02-25 http://www.mapscripting.com/how-to-use-geolocation-in-mobile-safari.html

02-26 https://paperplanes.world/

02-27 https://developers.google.com/web/showcase/2016/aliexpress

02-28 https://twitter.com/josephjames/status/779068864231505921

Chapter 3

03-01 https://adactio.com/journal/6246/

03-02 https://medium.com/@owencm/designing-great-uis-for-progressive-web-apps-dd38c1d20f7

03-03 https://css-tricks.com/snippets/css/system-font-stack/

03-04 https://drafts.csswg.org/css-fonts-4/#system-ui-def

03-05 https://caniuse.com/#search=system-ui

03-06 https://infinnie.github.io/blog/2017/systemui.html

03-07 https://superdevresources.com/material-design-web-ui-frameworks/

03-08 https://www.quora.com/Why-is-Google-Chrome-browser-named-as-Chrome/answer/Glen-Murphy

03-09 https://www.w3.org/TR/appmanifest/#display-modes

03-10 https://www.moovweb.com/anyone-use-social-sharing-buttons-mobile/

03-11 https://blog.easy-designs.net/archives/dont-sell-out-your-users

03-12 https://www.w3.org/TR/clipboard-apis/

03-13 https://wicg.github.io/web-share/

03-14 https://developers.google.com/web/updates/2016/09/navigator-share

03-15 https://hightide.earth

03-16 https://medium.com/@addyosmani/progressive-web-apps-with-react-js-part-2-page-load-performance-33b932d97cf2

03-17 https://cloudfour.com/thinks/why-does-the-washington-posts-progressive-web-app-increase-engagement-on-ios/

03-18 https://page-transitions.com/

03-19 https://css-tricks.com/native-like-animations-for-page-transitions-on-the-web/

03-20 https://developers.google.com/web/fundamentals/performance/rendering/stick-to-compositor-only-properties-and-manage-layer-count

03-21 http://barbajs.org

03-22 https://infrequently.org/2017/10/can-you-afford-it-real-world-web-performance-budgets/

03-23 https://developers.google.com/web/fundamentals/performance/prpl-pattern/

03-24 https://github.com/turbolinks/turbolinks

03-25 https://paul.kinlan.me/progressive-progressive-web-apps/

03-26 https://developer.apple.com/ios/human-interface-guidelines/overview/themes/

03-27 https://material-components-web.appspot.com/button.html

03-28 https://codepen.io/dsenneff/full/2c3e5bc86b372d5424b00edaf4990173/

03-29 https://csstriggers.com/

03-30 https://developers.google.com/web/fundamentals/performance/rendering/stick-to-compositor-only-properties-and-manage-layer-count

Chapter 4

04-01 https://thishereweb.com/understanding-the-manifest-for-web-app-3f6cd2b853d6

04-02 https://cloudfour.com/manifest.json

04-03 https://www.w3.org/TR/appmanifest/
04-04 https://manifest-validator.appspot.com/
04-05 https://developers.google.com/web/ilt/pwa/lab-auditing-with-lighthouse
04-06 https://infrequently.org/2016/09/what-exactly-makes-something-a-progressive-web-app/
04-07 https://www.youtube.com/watch?time_continue=831&v=m-sCdS0sQO8
04-08 https://developers.google.com/web/updates/2018/06/a2hs-updates
04-09 https://www.pwabuilder.com/
04-10 https://phonegap.com/
04-11 https://developers.google.com/web/updates/2017/10/using-twa
04-12 https://twitter.com/_developit/status/922512745378979840

Chapter 5

05-01 https://developers.google.com/web/tools/workbox/
05-02 https://material.money/
05-03 https://developers.google.com/web/fundamentals/instant-and-offline/offline-ux
05-04 https://www.thinkwithgoogle.com/intl/en-gb/consumer-insights/trivago-embrace-progressive-web-apps-as-the-future-of-mobile/
05-05 https://www.webcomponents.org/element/polymerelements/app-storage
05-06 https://wiki-offline.jakearchibald.com/
05-07 https://wicg.github.io/BackgroundSync/spec/
05-08 https://jakearchibald.github.io/isserviceworkerready/

Chapter 6

06-01 http://info.localytics.com/blog/the-inside-view-how-consumers-really-feel-about-push-notifications
06-02 http://info.localytics.com/blog/2015-the-year-that-push-notifications-grew-up
06-03 https://twitter.com/codevisuals/status/838881724016787457
06-04 https://youtu.be/PsgW-0M67TQ?t=22m50s
06-05 https://tests.peter.sh/notification-generator/
06-06 https://web-push-book.gauntface.com/
06-07 https://web-push-book.gauntface.com/chapter-05/02-display-a-notification/

Chapter 7

07-01 https://www.ampproject.org/

07-02 http://www.thesempost.com/google-search-amp-clicks-non-news-sites/

07-03 https://www.theverge.com/2016/12/6/13850230/fake-news-sites-google-search-facebook-instant-articles

07-04 http://ampletter.org/

07-05 http://blog.chartbeat.com/2018/02/15/google-is-up-what-to-do-about-it

07-06 https://cloudfour.com/thinks/autofill-what-web-devs-should-know-but-dont/

07-07 http://www1.janrain.com/rs/janrain/images/Industry-Research-Value-of-Social-Login-2013.pdf

07-08 https://developers.google.com/web/showcase/2016/ali-express-smart-lock

07-09 https://developers.google.com/web/showcase/2016/guardian-smart-lock

07-10 https://developer.mozilla.org/en-US/docs/Web/API/Web_Authentication_API

07-11 https://www.wired.com/story/webauthn-in-browsers/

07-12 http://www.adweek.com/digital/survey-56-of-u-s-consumers-have-abandoned-a-mobile-transaction/

07-13 https://baymard.com/lists/cart-abandonment-rate

07-14 https://developer.apple.com/documentation/applepayjs

07-15 https://lists.w3.org/Archives/Public/public-payments-wg/2016Jun/0013.html

07-16 https://github.com/GoogleChromeLabs/appr-wrapper

07-17 https://www.mobify.com/customers/pureformulas/

07-18 https://www.wompmobile.com/payment-request-api-case-study

07-19 https://www.youtube.com/watch?v=1-g1rvkORQ8&feature=youtu.be&t=6m21s

Chapter 8

08-01 https://articles.uie.com/kj_technique/

08-02 http://www.webpagetest.org/

08-03 https://webhint.io/

08-04 https://developers.google.com/web/updates/2018/05/first-input-delay

08-05 https://www.youtube.com/watch?v=yURLTwZ3ehk

Chapter 9

09-01 https://www.wired.com/2010/08/ff-webrip/

09-02 https://twitter.com/slightlylate/status/740228002311639040

09-03 https://twitter.com/brucel/status/740252470354542592

09-04 http://www.comscore.com/Insights/Presentations-and-Whitepapers/2016/The-2016-US-Mobile-App-Report

09-05 http://www.recode.net/2016/6/8/11883518/app-boom-over-snapchat-uber

09-06 https://twitter.com/scottjehl/status/740904147524947968

09-07 https://android-developers.googleblog.com/2016/05/android-instant-apps-evolving-apps.html

09-08 https://tech.economictimes.indiatimes.com/news/internet/for-flipkart-this-app-makes-rural-connect/59676200

09-09 https://developers.google.com/web/showcase/2017/ola

Resources

10-01 https://abookapart.com/products/going-offline

10-02 https://www.manning.com/books/progressive-web-apps

10-03 https://www.smashingmagazine.com/2016/02/making-a-service-worker/

10-04 https://web-push-book.gauntface.com/

10-05 https://webpushdemo.azurewebsites.net/

10-06 https://docs.google.com/document/d/1WNPIS_2F0eyDm5SS2E6LZ_75t-k6XtBSnR1xNjWJ_DPE/edit#heading=h.v5v9jr5n9i1w

10-07 https://developers.google.com/web/showcase/

10-08 https://medium.com/@addyosmani

10-09 https://mobiforge.com/design-development/bootstrapping-progressive-web-apps-with-amp-install-serviceworker

10-10 https://developers.google.com/web/fundamentals/security/credential-management/

10-11 https://www.smashingmagazine.com/2018/01/online-purchase-payment-request-api/

10-12 https://developers.google.com/web/tools/lighthouse/

10-13 https://whatdoesmysitecost.com

10-14 https://www.smashingmagazine.com/2017/03/world-wide-web-not-wealthy-western-web-part-1/

10-15 https://youtu.be/eG0ILA2k5qo

INDEX

A

Accelerated Mobile Pages (AMP)
    113-116
AJAX 84
animation 55-59
APIs
    backgroundSync 83
    Clipboard 37
    Credential Management 113-114
    Geolocation 22
    History 33-34
    Payment Request 113, 121
    Push 110
    Streams 52
    Web Authentication 120
    Web Share 39
app shell model 41
app stores 75-78
Archibald, Jake 91

B

banners and badges 68-74
Berriman, Frances 5-7, 144
Beverloo, Peter 110
browser chrome 32-34
browser support 24

C

caching strategies 81-88
Campbell-Moore, Owen 27
Chen, Andrew 14
Cloud Four 12, 124
comScore 14
credential management 117-120
CSS Working Group 28

D

display modes 35-36
domain sharding 133
Drasner, Sarah 42-43

E

engagement 20-21

F

federated login 117
feedback 53-54
F.I.R.E. 4, 7-8
First Input Delay (FID) 131
front-end features 135-136

G

Gaunt, Matt 110
Google Material Design 28
Graham, Geoff 27
graphical processor unit (GPU) 55-56

H

homescreen 61-62
HTTP/2 17
HTTPS 16-17, 132-133

J

JavaScript 48-51
    isomorphic 49
Jehl, Scott 141
JSON 48

K

Keith, Jeremy 7-8, 26, 82
KJ-Method 127

ABOUT A BOOK APART

We cover the emerging and essential topics in web design and development with style, clarity, and above all, brevity—because working designer-developers can't afford to waste time.

COLOPHON

The text is set in FF Yoga and its companion, FF Yoga Sans, both by Xavier Dupré. Headlines and cover are set in Titling Gothic by David Berlow.

This book was printed in the United States using FSC certified papers.

FSC®
www.fsc.org

ABOUT THE AUTHOR

Jason Grigsby is cofounder of Cloud Four, a small web consultancy with big aspirations. Since cofounding Cloud Four, he has had the good fortune to work on many fantastic projects, including the Obama '08 iOS app. He was founder and president of Mobile Portland, where he helped start the world's first community device lab. He is the coauthor of *Head First Mobile Web* (O'Reilly 2011) and one of the signatories of the Future Friendly web manifesto. He participated in the Responsive Images Community Group, which helped define the new web standard for responsive images. He tracks progressive web app case studies at pwastats.com. You can find him blogging at cloudfour.com; on his less frequently updated personal site, userfirstweb.com; and on Twitter as @grigs.

www.ingramcontent.com/pod-product-compliance
Lightning Source LLC
Chambersburg PA
CBHW040857210326
41597CB00029B/4877